# Discovering Dahlias

## A Guide to Growing and Arranging Magnificent Blooms

ERIN BENZAKEIN

*with* JILL JORGENSEN AND JULIE CHAI

PHOTOGRAPHS BY
CHRIS BENZAKEIN

Library of Congress Cataloging-in-Publication Data

Names: Benzakein, Erin, author. | Chai, Julie, author. |
Jorgensen, Jill, author.
Title: Floret Farm's discovering dahlias : a guide
to growing and arranging magnificent blooms /
Erin Benzakein with Jill Jorgensen and Julie Chai;
photographs by Chris Benzakein.
Description: San Francisco : Chronicle Books, [2020] |
Includes index. |
Identifiers: LCCN 2020024459 (print) | LCCN
2020024460 (ebook) | ISBN
9781452181752 (hardback) | ISBN 9781452181851
(ebook)
Subjects: LCSH: Dahlias. | Dahlias—Varieties. | Flower
gardening. | Flower arrangement.
Classification: LCC SB413.D13 B46 2020  (print) | LCC
SB413.D13  (ebook) | DDC 635.9/33983—dc23
LC record available at https://lccn.loc.gov/2020024459
LC ebook record available at https://lccn.loc.
gov/2020024460

MIX
Paper from
responsible sources
FSC
www.fsc.org    FSC™ C008047

Manufactured in China.
Design by Ashley Lima.

10 9 8 7 6 5 4 3

Chronicle books and gifts are available at special quantity
discounts to corporations, professional associations,
literacy programs, and other organizations. For details
and discount information, please contact our premiums
department at corporatesales@chroniclebooks.com or at
1-800-759-0190.

Chronicle Books LLC
680 Second Street
San Francisco, CA 94107
www.chroniclebooks.com

*For the fairies.*

# CONTENTS

# FALLING FOR DAHLIAS

Of all the flowers I've ever grown, dahlias are my favorite. These treasures are one of the most well-loved and widely grown flowers for cutting because they come in a dazzling rainbow of colors, they produce an abundance of flowers from midsummer into autumn, and the range of shapes and sizes available is staggering. In addition to their being such a beloved cut flower, their incredible ability to multiply over the course of a growing season is unmatched: you can start with a single tuber or rooted cutting, and by season's end be digging a full clump that contains 3 to 10 babies from the original plant. Similar to a sourdough starter, once you have it and as long as you take care of it, you'll have a steady supply to share with others and plant for yourself, every season.

Many years ago, when I was just starting to grow cut flowers, I got a phone call from a local flower grower telling me to load up the kids and my shovel and head over to her house. It was a crisp morning in October, just after our first autumn frost, and I found her digging up dahlias. At the time I knew very little about these bloomers, only that I admired them every time I visited her garden. We worked through the morning lifting her massive clumps of tubers, splitting off a chunk of each variety for me to take home to my garden. Her generosity was my first real taste of just how giving and passionate flower people are. When I offered to pay her or pull weeds in exchange for the station wagon full of tubers that she had so generously

9

shared, she said she didn't want anything in return. Her one request was that I pass along some of the abundance to another gardener in need once my plants were established.

Since that fateful October day many years ago, my dahlia garden has grown beyond anything I ever imagined. This past season we grew nearly 800 unique varieties and over 18,000 plants in total, a far cry from a station wagon full of muddy tubers. One thing I've learned when it comes to dahlias is that once you've been bitten by the bug, there's no going back. They have a strangely magical quality that somehow ends up taking over your life in the most fun and beautiful way. But the best part of all is having the opportunity at the end of each season to pass their magic on to others.

While dahlias have a large and passionate fan club, with growers spanning the globe, there has been a notable shortage of current information about how to cultivate these beloved plants. On the pages ahead, I share how to get started with dahlias; demystify their many sizes and forms; teach you step by step how to grow, harvest, and arrange these spectacular flowers; and feature hundreds of my favorite varieties, sorted by color.

I have been blessed by the generosity and optimism of so many gardeners along the way and offer this book as a gift in return. My hope is that it will inspire you to fill your life with dahlias, and even more importantly, share them with others.

CHAPTER ONE

# UNDERSTANDING DAHLIAS

Part of what makes dahlias so beloved is their spectacular diversity. These glorious flowers come in a huge range of sizes, forms, and colors, offering a variety for virtually every garden space and personal preference.

As you learn more about dahlias, you'll find that different organizations classify dahlias in different ways, usually based on color, bloom size, and flower form. For example, the American Dahlia Society (ADS), a widely known dahlia-focused group, recognized 6 sizes, 20 forms, and 17 colors at the time of this printing, and these numbers can change as new dahlias are discovered. The ADS then combines these in a detailed system in which they place each variety. When shopping, you'll find that some dahlia sellers follow the ADS system, while others make no mention of it at all. Since I created this book for those who want to grow and design with these incredible flowers, I've pared down the classification information so it's as straightforward and easy to understand as possible. The details ahead are based on what I've found that most home gardeners, flower farmers, and floral designers need to know and will be most useful in finding the dahlias that are right for you.

POMPON (P)  MINI BALL (MB)  MINIATURE (M)  BALL (BA)  SMALL (BB)

# SIZES

Whether you're growing to make arrangements or simply to enjoy dahlias in your yard, there is a huge range of sizes to choose from. While some sources approach size classifications in different ways, we find the following 8 sizes to be the most commonly used for general gardening and designing.

### POMPON, UP TO 2 IN (5 CM), ALSO KNOWN AS P

Miniature round blooms look like little lollipops on long, wiry stems. They're a fantastic addition to any bouquet, as they're long lasting and weather resistant. Plants are typically on the smaller side.

### MINI BALL, 2 TO 3½ IN (5 TO 9 CM), ALSO KNOWN AS MB

These have the same rounded shape as ball varieties but are very easy to incorporate into arrangements because of their versatile size. They're also long lasting in bouquets and weather resistant.

### MINIATURE, UP TO 4 IN (10 CM), ALSO KNOWN AS M

This diverse class features many petite blooms that are well loved by all. So many unique forms fall into this range of sizes.

### BALL, OVER 3½ IN (9 CM), ALSO KNOWN AS BA

They're the largest of the rounded blooms and come in a wide range of choices. They make excellent cut flowers as they're some of the longest lasting, can handle heat, and are generally weather resistant.

MEDIUM (B)

LARGE (A)

GIANT (AA)

### SMALL, 4 TO 6 IN (10 TO 15 CM), ALSO KNOWN AS BB

This class boasts the greatest number of varieties by far. They're well suited for home gardens, and flower arrangers love them because the blooms are easy to incorporate, as they're not too big or fragile.

### MEDIUM, 6 TO 8 IN (15 TO 20 CM), ALSO KNOWN AS B

One of the biggest, most versatile, and best represented sizes, this broad group includes many of the most coveted varieties for flower arranging.

### LARGE, 8 TO 10 IN (20 TO 25 CM), ALSO KNOWN AS A

The sizable blooms in this class make wonderful additions to bigger flower arrangements and look spectacular in the garden. They're slightly smaller than dinner plate dahlias, but still have high impact and are easier to use in arrangements.

### GIANT, 10+ IN (25+ CM), ALSO KNOWN AS AA

These are often referred to as dinner plate dahlias. The massive blooms in this class require a little extra protection from weather due to their tremendous size and delicate nature. Well suited for large-scale arrangements, these beauties are real showstoppers.

INFORMAL DECORATIVE (ID)

FORMAL DECORATIVE (FD)

STELLAR (ST)

PEONY (PE)

BALL (BA)

MINIATURE
BALL (MB)

POMPON (P)

18

ORCHID (O)

ORCHETTE (OT)

SINGLE (S)

MIGNON SINGLE
(MS)

# FORMS

Along with their range of sizes, dahlias come in a dazzling array of forms, most of which fall into one of the following groups. Some are perfectly symmetrical, with tightly spaced petals, while others have a loose, lush look. They may have different petal styles and lengths, with open centers or not. When it comes to choosing varieties, people's preferences vary widely—some prefer cactus types, while others will grow only informal decoratives. I'm personally drawn to the oddballs, including anemone, stellar,

**SEMI-CACTUS (SC)**

**STRAIGHT CACTUS (C)**

**INCURVED CACTUS (IC)**

**LACINIATED (LC)**

**WATERLILY (WL)**

**ANEMONE (AN)**

**COLLARETTE (CO)**

**NOVELTY OPEN (NO)**

**NOVELTY
FULLY DOUBLE (NX)**

incurved cactus, and orchette. Over time you might find yourself gravitating to particular classes, and understanding the differences between them will help you when selecting new varieties. I offer detail about each form on the following pages.

*continued* ⟶

### INFORMAL DECORATIVE (ID)

Informal decoratives have a soft, romantic quality to the blooms, and their petals are typically lush and billowy. This class includes many of the most beautiful and useful varieties for making arrangements.

### MINIATURE BALL (MB)

Similar to ball varieties, these include a wide range of smaller, rounded blooms that are very long lasting and weather resistant. If you're looking for hardworking varieties to sell, these fit the bill.

### FORMAL DECORATIVE (FD)

Though there are fewer overall varieties of this kind, formal decoratives are among my favorites to use for designing. The large, showy blooms have a more uniform and refined quality than some other types, and they are a stunning addition to large arrangements.

### POMPON (P)

These are some of the cutest varieties imaginable. The petite ball-shaped flowers are the size of large shooter marbles, ideal for adding interest and texture to bouquets. Floral designers can't get enough of them.

### BALL (BA)

With the longest-lasting, most weather-resistant, medium-size round blooms, varieties in this class are essential to add to your beds if you're growing dahlias for market.

### LACINIATED (LC)

This class is filled with shaggy, fluffy blooms with outer petals that look as if they've been snipped with pinking shears. Flowers are more prone to weather damage, since they are on the delicate side, but if given a little extra protection, they are wonderful textural additions to the garden and the vase.

### NOVELTY FULLY DOUBLE (NX)

This class is another grab bag of varieties that don't easily fall into other defined classes. Their common trait is a closed, tight center and petals that are proportionate in size.

*continued* ⟶

FORMAL DECORATIVE
(FD)

NOVELTY
FULLY DOUBLE (NX)

BALL (BA)

INFORMAL
DECORATIVE
(ID)

MINIATURE BALL (MB)

POMPON (P)

LACINIATED (LC)

INCURVED CACTUS (IC)

STRAIGHT CACTUS (C)

STELLAR (ST)

WATERLILY (WL)

SEMI-CACTUS (SC)

NOVELTY OPEN (NO)

ANEMONE (AN)

## WATERLILY (WL)

One of the classes most loved by flower arrangers, waterlily types have flowers borne on long, strong stems and typically hold their heads upright. Aptly named, the tidy saucer-shaped blooms resemble floating waterlilies.

## ANEMONE (AN)

This class includes fun novelty varieties that most people would never guess are dahlias because of their unique appearance. Flowers have a ring of reflexed petals that surround a domed pincushion-shaped center, similar to double-flowered echinacea.

## STRAIGHT CACTUS (C)

Brimming with spiky, textural varieties, this class has had dahlia breeders' focus for many years because the flowers are so striking in the garden. I've found these porcupine-like blooms to be quite fragile when it comes to weather damage, and their tips tend to show heat stress very easily.

## SEMI-CACTUS (SC)

Another popular class with breeders, semi-cactus types come in loads of choices in every color. They are similar in shape to straight cactus varieties, but with petals that are less defined and tubular and much more relaxed. These flowers typically have long stems and are good for flower arranging.

## INCURVED CACTUS (IC)

Of all the cactus types, this class is by far my favorite. Their flowers remind me of Muppet characters with their twisted, tubular, feather-like petals. They are a fun addition to the garden and are always a topic of conversation with visitors.

## NOVELTY OPEN (NO)

This class is somewhat of a catchall for rare varieties that don't clearly fit into other classes. What makes this group unique is that the flowers' single centers are proportionate to the ring of outer petals.

## STELLAR (ST)

This class is filled with striking, textural blooms that are very eye-catching. Their sharply edged petals are reflexed back toward the stem, resembling colorful shooting stars.

23

*continued* $\longrightarrow$

## ORCHETTE (OT)

This class is similar to orchid types but has flowers that boast a fluffy collar of inner petals, giving the star-shaped blooms a more delicate, romantic appearance. I love incorporating them into bouquets because they add a fun, textural quality.

## COLLARETTE (CO)

Darling single blooms that have a collar of ruffled petals encircling glowing single centers make varieties in this class lovely as bouquet additions.

## SINGLE (S)

This class is one of my favorites because the daisy-like flowers are a cheerful addition to arrangements. While petal shapes vary widely from rounded to pointed, they all have a striking single center.

## ORCHID (O)

Fast becoming a favorite among designers, this class is filled with long-stemmed, star-like flowers with petals that roll inward. Pollinators love them, they're an eye-catching addition to the garden, and these beauties are also perfect for bouquets.

## MIGNON SINGLE (MS)

With miniature single blooms that have rounded petal tips, these petite flowers have a diameter less than 2 in (5 cm).

## PEONY (PE)

This class has few varieties. Flowers have a ring of at least 2, but not more than 5, rows of petals around the outside edge encircling a single, open center.

MIGNON SINGLE (MS)

COLLARETTE (CO)

PEONY (PE)

SINGLE (S)

ORCHETTE (OT)

ORCHID (O)

# COLOR

Dahlias come in bright, saturated shades as well as subdued, muted ones. A single bloom might include multiple colors that are next to each other on the color wheel for a look that's naturally ombre, or may have shades that are further apart, like white and plum, that offer greater contrast. Since we've found that most designers and gardeners we know choose plants based on the color palette they're working with, we've arranged our Variety Finder on page 137 by color so you can discover the varieties that suit your needs and personal style.

27

CHAPTER TWO

# GROWING AND CARE

Of all the flowers you can grow, dahlias are some of the least demanding and most rewarding. When given the right conditions and care, they'll repay you with buckets of blooms for months, and their tubers will multiply abundantly underground so you'll have more planting stock with each passing year.

When I first started growing dahlias, I made every mistake in the book, from planting way too early in the spring and losing my young plants to a late cold snap, to not providing adequate support early enough and having thousands of the most beautiful plants knocked over and broken during a freak windstorm in early summer. I've also lost more tubers than I can count to improper dividing and poor storage practices. While each of these mistakes was devastating at the time, I've learned so much about this incredible group of plants along the way.

This section takes you through planting time in spring, harvesting in summer and beyond, and digging, dividing, and storing in autumn and early winter. Depending on who you ask, every dahlia grower will have their own preferred methods, but the guidance that follows is based on what's been most successful here on our farm.

# HOW TO GROW

Dahlias are relatively easy to grow and need only a few essentials: good soil, adequate water, and lots of sun. These hardworking beauties are quite sensitive to the cold, and if you live in an area with cold winters you'll need to dig up your tubers in autumn and store them in a frost-free place until it's time to replant in the spring.

## SUN

Dahlias thrive in warm sunny weather. Plant them in a spot that gets at least 6 hours of direct sunlight each day. Otherwise your plants will get leggy reaching for the light, and they won't bloom as abundantly as possible.

## FERTILE SOIL

One of the keys to success with dahlias is planting them in healthy soil. Even if you're not blessed with great ground, it's possible to transform a less-than-ideal plot in a short amount of time. Start by taking a soil sample and having it tested by a soil lab, which is an inexpensive way to find out what nutrients might be missing. Each lab will have instructions on how to take and send your sample, but it generally involves digging down about 1 ft (30 cm), then gathering a small amount of soil from multiple locations around your garden until you have a quart-size (liter-size) sample.

Regional agricultural extension offices can often recommend a soil testing lab in your area. If you don't have a local option, you can mail soil samples to a lab farther away (see Resources, page 219). When you have your samples analyzed, you'll get a detailed report on your garden's soil, including what percentage of organic matter is present, what trace minerals are lacking, and what type of amendments—such as compost, bone meal, lime, kelp, and rock phosphate—and how much of each you should add to remedy any problems. Be sure to let the testing facility know whether you plan on growing organically so they can recommend natural products instead of ones containing synthetic chemicals.

Even though taking a soil test can feel like a nuisance and requires a few additional steps and time before you can start planting, the extra effort on the front end will be well worth it. I've seen so many gardeners who've struggled with plant health and disease issues, and who have completely turned things around in a single growing season after they tested their soil and made the recommended adjustments. If you plan to grow organically, getting a soil test is a must.

After making any necessary changes from your soil test, prep your beds for planting. Dahlias do best when given plenty of organic matter, such as compost, and a generous dose of a balanced, natural fertilizer. I recommend spreading 2 to 3 in (5 to 8 cm) of high-quality compost over your beds, and then broadcasting a general fertilizer on top of the compost at a rate of approximately 10 lb (4.5 kg) per 1,000 square ft (93 square m). Mix the compost and fertilizer into the top layer of the soil using a walk-behind rototiller. If you don't have a tiller, simply use a shovel or pitchfork to work the amendments evenly into the soil. Whether you grow your dahlias in rows or incorporate them into your existing landscape, the soil prep is the same.

It's also important that your soil drains well, since standing water, or wet clay that holds moisture over extended periods of time, can cause tubers to rot. If you don't have a spot that's freely draining, you can create raised beds so that excess water will drain away.

## SPACE

If you're growing dahlias in your landscape among other plants, they'll require a bit more space than those in rows—give each dahlia a minimum of 3 ft (91 cm) on each side to spread out. If you're growing them in rows, I suggest planting 2 rows per 3 ft (91 cm) wide bed, spacing the rows 18 in (46 cm) apart with 12 in (30 cm) between plants. This is the spacing we use at our farm for every size of dahlia, and while it may seem tighter than you're used to, the plants will do just fine if you harvest from them regularly and provide plenty of support.

## GROWING DAHLIAS IN SMALL SPACES

Even if you don't have a large yard in which to grow your dahlias, you can still fit in quite a few of these hardworking plants if you exercise your creativity. I've seen gardeners who took advantage of unused spaces and built raised planter boxes on top of an abandoned tennis court, planted in the strip between the sidewalk and the street, and grew in a narrow planter tucked up against the garage in an alley. There are so many places you can sneak in a few plants if you get creative.

33

If you don't have any actual ground to grow in, you can plant dahlias in large pots. To give them enough room to spread out and thrive, I recommend growing in half wine barrels or large galvanized tubs—or a container equivalent in size—that are at least 1 ft (30 cm) deep and 2 ft (61 cm) wide. If you're growing dahlias in pots, be sure to choose knee-high varieties that top out at 3 ft (91 cm) tall, such as 'Amber Queen', 'Totally Tangerine', and 'Waltzing Mathilda'.

Keep in mind that when you grow any plant in a container, it will require a lot more care than those growing directly in the ground. In addition to needing regular deep watering, especially during the height of summer, container-grown plants need to be fertilized monthly with a balanced organic liquid fertilizer.

### PLANTING

Dahlias are cold sensitive, so plant them outside only after all danger of frost has passed in spring—for us, in Washington's Skagit Valley, this is late April to early May.

Tubers produce exact clones of their mother plants, so most people grow dahlias from tubers, since they want specific varieties. Depending on how many tubers you're planting, you can dig either a long trench or individual holes, and set tubers 4 to 6 in (10 to 15 cm) deep. Place each tuber horizontally with its growing eye (if visible) facing up, then cover with soil. Be sure to label as you plant, because it can be very easy to forget what you planted in a given spot once you bury the tubers. On the farm, we place a stake at the beginning of each row containing a single variety. When planting single tubers in the landscape, we label plants individually.

If planting from rooted cuttings (see How to Take Cuttings, page 79), which are also clones of the mother dahlia, plant outdoors like any other annual plant with the foliage above the soil line. Because cuttings are very tender, be sure to wait until after the weather has warmed and any threat of frost has passed—in Washington, this is around Mother's Day.

To grow from seed, see How to Grow from Seed, page 85.

### WATER

Start watering dahlias after the first green shoots appear above the soil, which typically happens about a month after planting. Overwatering before sprouts appear can lead to tuber rot. It's fine if it rains, but don't start any other kind of watering until you spot active growth.

Once plants are visible, but before they've gotten too big, make sure you have some kind of irrigation in place. If you're planting in an existing landscape, soaker hoses are perfect; if you're growing in rows, I recommend drip irrigation because it is much more affordable. Dahlias need a lot of water during the growing season, so we run two or three lines of drip irrigation per 3 ft (91 cm) wide bed, using T-tape that has perforations every 8 in (20 cm).

To help conserve moisture, we add a thick layer of mulch around young plants once they've emerged from the soil. Shredded leaves, straw, or dried grass clippings are all great choices. Mulching is not necessary, but does help cut down on weeding—a chore that can be difficult and time consuming if you let weeds mature—and holds the moisture in the soil, especially in the heat of summer, when dahlias will need a steady supply of water. Once plants are actively growing, we water deeply once a week. In the heat of summer, we increase that to two to three times per week depending on the weather. Your watering needs will vary based on where you live, your weather, and your soil.

For long, strong stems, the practice of pinching is one of the most important techniques to know. This encourages plants to produce more branches near the base, which increases the total number of flowering stems per plant and encourages longer stem length. Once plants are between 8 and 12 in (20 to 30 cm) tall, use sharp pruners to snip the top 3 to 4 in (7 to 10 cm) off the plant, just above a set of leaves. This causes the plant to send up multiple stems from below the cut, resulting in more abundant flower production. Pinching can feel counterintuitive, since you're removing what would be a flowering stem, but it will result in many more usable flower stems over the course of the season. It also helps prevent the plant from developing large, hollow stems, which can often become as big as broom handles and are impossible to arrange with.

## STAKING

Plants grown in the right conditions and with the care previously described will inevitably grow tall and heavy. Lush, healthy dahlias require sturdy staking to stay upright, and it's important to have it in place before plants grow too large and topple over from the weight of their showy blossoms.

If you're growing dahlias in your garden, you can pound individual stakes into the ground next to the tuber so you don't disturb roots later on. As plants grow, use twine or sisal string to tie them to the stakes, at every 12 to 18 in (30 to 46 cm) of height, to support the growth.

If planting dahlias in long rows, corralling them is by far the most efficient and easy-to-install support method. When plants are 18 to 24 in (46 to 61 cm) tall, we pound a 4 to 5 ft (1.2 to 1.5 m) metal T-post every 10 ft (3 m) along the outside edge of the beds, then run a layer of baling twine around the entire perimeter, looping it around each T-post and pulling it tight as we go in order to create an enclosed box around the growing bed. If the string is not pulled tight, it will sag and droop, so be sure to hold tension on the twine as you go so that it remains taut. Tie the first layer of twine about 2 ft (61 cm) off the ground, and the second layer about 18 in (46 cm) above that. This double layer of twine will give plants all the support that they need to stay upright during the growing season, and you can add additional layers of twine as needed.

## GROWING DAHLIAS
## UNDER COVER

When we have big, fluffy, fragile blooms—like 'Café au Lait' and others in the dinner plate category—that are destined for market or show, we grow them under cover in unheated hoop houses. This strategy has dramatically increased productivity, stem length, and flower quality, since the heavy blooms are protected from wind and rain.

Because the hoop houses warm up earlier than the outdoors in spring, we generally plant tubers inside (in the ground) 4 to 5 weeks ahead of those going into the field. Inside the cozy, sheltered space, our dahlias often reach 7 to 8 ft (2.1 to 2.4 m) tall. In the hoop houses, we use the same corralling method mentioned earlier but instead use taller 6 ft (1.8 m) T-posts and an additional layer of twine.

While growing under cover improves dahlias' flower quality, there are a few drawbacks to keep in mind. Because the environment is warm and humid, insects, such as spider mites, and disease, such as powdery mildew, are more prevalent. So be sure to make the paths between rows a bit wider so that you give your dahlias adequate air flow, which can help reduce pests and diseases, and so you have enough space to walk between plants.

Know that growing in a hoop house or other protective structure is not essential, and dinner plate varieties can thrive in the garden and in the field. But if you want to increase your likelihood of raising perfect blooms that command top price, I highly recommend growing under cover.

## DISEASES

Like any plants, dahlias are prone to bacterial, viral, and fungal diseases when stressed. The following are the main types.

• The most common bacterial disease in dahlias is crown gall, which is identified by bumpy, cauliflower-like growths around the neck of the tubers, or large growths on the tuber clump. There is no treatment for crown gall; remove and throw away infected plants and tubers, otherwise the bacteria will spread to your other plants. Disinfect any garden tools that come in contact with infected plant parts, using a solution that's 10 percent bleach and 90 percent water, to prevent spreading the disease.

• Numerous viral diseases impact dahlias, and the telltale signs that a plant has a virus are yellow streaking or spots on the leaves and veins, and stunted plant growth. Viruses live in both the plant and the tuber, and there is no known treatment or cure for them, so regular monitoring and removal of infected plants is the best course of action. As soon as a virus is identified, pull out and destroy all parts of the plant immediately. Do not compost any parts, to prevent spreading the virus further.

• Fungal diseases such as powdery mildew, botrytis, leaf spot, and smut are spread by airborne spores. The best way to prevent them is by giving plants good airflow and proper care and watering as previously outlined, and by keeping the garden clean of any diseased debris.

## INSECTS

The kinds of insects and pests that might harm your dahlias will vary based on a range of environmental factors, the time of year, and where you live. In every case, the healthier you keep your plants, the less susceptible they'll be to insect damage, so we do a lot of work up front and throughout the growing season to keep our plants at their strongest.

Keep in mind that whatever treatments you use in your soil and apply to your plants will very likely come in contact with you, your children, and your pets at some point, so I advise using organic methods as much as possible. Growing naturally takes more effort and attention, but pays off in the quality of the flowers and in your and your family's health. While growing the perfect blooms is always exciting, if it requires exposing yourself and your loved ones to toxic chemicals, I would personally rather have a few more bug-eaten flowers and know my garden is a safe place to be.

The following are some of the most common pests that you might see on your dahlias.

• Slugs and snails are one of the biggest threats to young dahlias because they come out at night and eat the tender new growth right at ground level. The best way to combat these slippery buggers is to put down bait about 2 weeks after planting or as soon as you see sprouts starting to emerge from the soil. I use Sluggo, an organic option that is safe for both children and pets.

• Earwigs are my personal nemesis; they always seem to be lurking, ready to chomp the most perfect blooms. Sluggo Plus, a certified organic product that is safe around children and pets, works wonders on these destructive little beasts.

• When we've had extreme cases of aphids, we've resorted to using a backpack sprayer to apply insecticidal soap to buds and leaves during the coolest parts of the day. There are a number of organically approved brands that you can find at most garden centers.

39

• Western flower thrips have become problematic throughout the United States. I've found that they tend to feed more heavily on flowers from plants that are stressed, especially by drought, and they are most obvious on light-colored blooms. To combat them, I recommend using beneficial nematodes, specifically *Steinernema feltiae*. These are tiny little worms that occur naturally in soil, but when concentrated and applied directly to the soil can be an effective organic solution to these pests.

• As Japanese beetles are migrating across the United States, many growers are having to employ extreme practices to protect their dahlias from these hungry predators. The most effective organic approach is covering individual buds with organza bags and securely tying them around the stems so the beetles can't reach the flowers. Growers harvest blooms when they are about one-half to three-quarters open, then bring the flowers inside and remove the bags. Even though this process is time consuming, many flower farmers have found it to be successful on a large scale.

You may also encounter other pests that are specific to your region. For advice on how to identify and combat them, contact your local Master Gardeners group or county extension office.

*PHOTO BELOW:* **APHIDS**

*OPPOSITE PAGE, CLOCKWISE FROM TOP LEFT:* **CROWN GALL, POWDERY MILDEW, SNAIL, VIRAL DISEASE, ORGANZA BAGS**

# HARVESTING

While dahlias are not a particularly long-lasting cut flower—you can expect them to look good for about 5 days in the vase—their brilliant, colorful blooms make up for their fleeting existence. The tricks for getting the most out of these beauties are harvesting them at the proper stage and keeping them well hydrated afterward. Follow these steps to get the longest life from your dahlias.

**Harvest at the right stage and time of day.** Since dahlias don't unfurl much once they've been harvested, it's important to cut them when they're almost fully open (with a few exceptions that are mentioned in the Variety Finder on page 137), but at the same time not overly ripe. Check the back of each flower head, looking for firm and lush petals; papery or slightly dehydrated ones signal old age, and these blooms will shatter and drop their petals soon after you pick them. Because dahlias are delicate, you'll want to harvest them during the coolest hours of the day, either morning or evening, when plants are the most plump and hydrated, since they will recover more easily from the shock of being cut. Always take a water-filled bucket with you so you can immediately place stems in water while you're working.

**Cut long stems.** Longer stems are better for arranging and command a higher price if you plan on selling them, so when cutting, aim for a stem length of at least 12 to 15 in (30 to 38 cm). This encourages the plant to send out more branches at the base, resulting in longer stems and more of them. If you are timid and cut only short 6 to 8 in (15 to 20 cm) stems, over time you will find that subsequent flowers are borne on shorter, weaker stems that are harder to use in bouquets.

Many gardeners are afraid of cutting too deeply and removing too much of the plant; they don't want to sacrifice any of the flower buds. But I've found that the plants we grow and harvest very aggressively produce more flowers with longer stems over the course of a season than those not picked heavily. Like pinching, cutting deeply can feel counterintuitive and like you're harming the plant, but don't worry: this approach will reward you with an abundance of long-stemmed flowers.

**Pick regularly.** If you're growing dahlias for cut flowers on a larger scale, it's much more efficient to harvest at regular intervals than it is to examine each individual flower head for the proper stage of ripeness. On our farm, we harvest from our dahlia patch every 3 days, rain or shine, and never leave a single ripe flower on the plants. This protocol ensures that come harvest day, every flower is at the perfect stage and no blooms are left to fade and go to seed. In the early years when we would skip a scheduled harvest day, we ended up spending twice the amount of time in the patch sorting through flowers and culling those that were past their prime. The 3-day rule was a game changer, allowing our harvest crew to comb the patch quickly and efficiently, since every bloom was at the perfect stage.

**Deadhead.** Unless you're leaving seedpods to mature on the plants for breeding purposes, be sure to remove any spent blooms so that the plants continue to put energy into flower production rather than making seeds. This practice is called deadheading and is an important ritual in the cutting garden if you want a steady stream of beautiful blooms for the longest amount of time.

**Remove lower leaves.** After harvesting your dahlias, remove any leaves from the lower half of stems. This helps in two ways: One, it minimizes wilting, since there is less foliage to rehydrate, and two, it helps flowers drink, since leaves that are submerged in water will start to decay very quickly, encouraging bacteria that prevent stem ends from taking up water.

**Let stems recover.** If you harvest at the right time of day and your dahlias are well hydrated, you can arrange with them right away. But if they're at all wilted, it's important to give your freshly harvested blooms time to rest, prior to arranging, so that the stems have a chance to drink after the initial shock of being cut. Depending on how many flowers you're working with, you can hydrate them in one of two ways. If you plan to sell your dahlias, add hydration solution (two I like are Hydraflor from Floralife or OVB from Chrysal) to your buckets. This will extend the vase life of your dahlias by an additional 2 to 3 days. If you're growing dahlias purely for personal enjoyment and need them to rebound quickly, you can use the boiling water trick: After harvesting, dip their stem ends into boiling water for 7 to 10 seconds, at which point you will notice the stem ends changing color. Then place them in a vase or bucket of cool water. Dahlias benefit greatly from being stored in a cool environment, so once they're in water, I let them rest in a shady, protected spot for at least 3 to 4 hours. If you have access to a floral cooler, that's even better. We regularly keep freshly cut stems in a 38°F to 40°F (3.3°C to 4.4°C) cooler for 2 to 3 days before delivery, with great success. If you don't have a cooler, not to worry. Simply keep freshly cut blooms in a cool place such as a garage or basement away from direct sunlight.

**Use flower food.** When you're ready to arrange, add flower food to your vase to keep the water clear and the flowers more vibrant and long lasting. If you don't have flower food, be sure to change your vase water every other day. Freshly cut dahlias last longer when kept away from heat, bright light, and ripening fruit (such as a bowl of fruit on the table), since the ethylene emitted by ripening fruit and vegetables can cause flowers to wilt faster.

# CHAPTER THREE

# DIGGING, DIVIDING, AND STORING

The first autumn frost brings dahlia season to a screeching halt. It's always a bittersweet time for me as I watch the field go from a sea of blooms to blackened plants overnight. And though I'm ready for a slower pace following months of hard, focused work on the farm, having to say goodbye to these beauties is never easy. But before we can close the garden down for our coldest time of year, we have to complete our last big job: digging up all of our dahlia tubers and storing them safely for the winter.

Dahlia digging is one of my least favorite tasks of the year. After a hard frost, the once-lush green plants melt down into a pile of slimy foliage that starts to rot as soon as it thaws. Stepping into the decaying field is always a little daunting, but once we start cutting down the faded plants and removing them from the growing beds, excitement kicks in. It's a thrill to imagine the thousands of tuber clumps lying just below the soil surface.

When you have a lot of dahlias to dig up in the autumn, it's always so much easier and more fun if you have a team to take on the task together. Before we had a farm crew, I would invite friends and neighbors over to lend a hand with the big job in exchange for some clumps to grow in their own gardens. With enough help, even the biggest dahlia patch can be cut down, cleaned up, and dug in no time.

# COLD-WEATHER CARE

### TIME IT RIGHT

It's important to let tubers cure properly before you dig them up, so that they last in storage. If you pull them up while they are still actively growing or immediately after the first frost, the skin on the tubers won't have had enough time to toughen, so I recommend that you wait at least 10 to 14 days after your first hard frost (one that will kill all the warm-season annuals in your garden) before digging. We even wait a bit longer to start the process, after a few hard frosts, which is typically early November in Washington. In the past, I've jumped the gun and found that tubers dug without proper curing have a much higher rate of shriveling in storage and seem to be more susceptible to rot.

While tubers need to properly cure, you want to make sure and get them out of the ground before it freezes solid. In most climates, there is a brief window to tackle this project, and it's important that you don't delay. In cold climates, if you miss this safe digging window your clumps will freeze in the ground and won't survive.

### PREP, THEN DIG

Before digging up tubers, remove all of the staking and irrigation materials from your growing beds so you can get to the plants easily. Once the beds are clear, cut the blackened plants to the ground, remove all of the debris, and start digging. If you live in an area that doesn't freeze, you can trigger your tubers to

cure by cutting your plants down to the ground and letting them sit for about 2 weeks before digging. In cooler climates, this method can be used if you need to dig your dahlias before cold weather arrives. The best tool for digging is a pitchfork instead of a shovel. I use a 4-tine pitchfork, since it allows the dirt to easily fall away from the dahlia clumps and there's less likelihood of damaging the tubers as you might if you sliced through them with a shovel. When digging, insert the pitchfork at least 1 ft (30 cm) away from where center stalks were, and rock your pitchfork back, carefully removing the tubers from the soil. Take care when pulling clumps out of the ground, since they can be fragile and break easily.

Then shake off the excess dirt from the clumps, taking care not to damage them. It can be tempting to drop the clumps on the ground or shake vigorously to remove the dirt, but this will break many of the tubers' necks, leaving them susceptible to rot and therefore useless.

## DECIDE WHEN TO DIVIDE

Depending on how much time you have after you've dug up your dahlias, you can either wash and divide the tubers right away or store them in a cool, frost-free place until you have time to tackle dividing, a process detailed on page 61.

Many smaller growers dig and divide over consecutive days before storing. I, and other flower farmers I know, dig and store right away, then divide in winter (typically January and February) so we have year-round work for our crews. Home gardeners who have adequate storage space, are growing smaller quantities, and don't have the pressure of needing an exact count or processing a lot of volume often dig and store right away, then divide closer to planting time in spring.

If you'll be storing your clumps for any length of time before dividing, be sure to leave a light layer of soil on the clumps rather than washing them right away, because the soil will help hold in enough moisture to prevent shriveling in storage. Also note that if you live in a dry climate, such as much of California or the desert, you'll want to divide and/or store your dahlias within 48 hours of digging. Otherwise they'll dry out and won't be viable.

## LABEL CLEARLY

It's essential to label your tubers before you store them so that, when planting time rolls around, you'll know what you have.

For single clumps of a given variety, I cut a length of brightly colored flagging tape, write the variety name on one end, and securely tie it around the remaining stalk tip so that it is easy to see.

If you have several clumps of a variety, it's much easier and faster to label the container that you'll store them in rather than labeling individual clumps. My favorite way to store tubers before dividing is in black plastic bulb crates. You can get these from large flower farms and/or nurseries that grow a lot of bulbs. However, if you can't get these, milk crates work well because they allow excess dirt and moisture to escape and are easy to pick up by their handles. We double label all of our bulk tubers by tying a piece of flagging tape with the variety name on it through the handle of each crate and then adding a wooden stake with the variety name on top of the clumps in the bin. This system has saved us many times, since rodents seem to love nibbling on labels of any kind.

## OVERWINTERING DAHLIAS

Growers in mild climates, such as USDA zone 8A or higher, which has an average low temperature of 10°F to 15°F (-12°C to -9°C), can successfully overwinter their dahlias—which means leaving them in the ground through the cold months—if given adequate winter insulation. These plants will flower much earlier, oftentimes 4 to 6 weeks ahead of tubers and cuttings planted out in the spring.

To overwinter, cut all foliage to the ground, then spread at least 1 ft (30 cm) of insulating mulch, such as straw or leaves, over the plants. Many of the growers I know in wetter climates go even further by covering their mulched plants with plastic or landscape fabric to keep the clumps on the drier side over winter. One thing to note, if you have voles, is that they like to nest in the cozy mulch. You might find them eating some of your tubers, but the trade-off might be worth it because you can bypass the time-consuming process of digging, dividing, and storing.

If you overwinter your dahlias, be sure to pull the mulch away from the crowns as soon as new growth appears in the spring. In my experience, insects are a bit more of a problem for overwintered plants because slugs, snails, and earwigs can easily hide in the bedding. So, once the crown is exposed, immediately put down some bait to combat the pests. I recommend Sluggo Plus, which is effective and safe around children and pets.

The biggest drawback to overwintering your dahlias in the ground is that by the second year, the clumps will often be massive and, should you decide to divide them at a later date, they will be pretty difficult to get out of the ground. This low-maintenance approach also comes with added risk—if you have an uncharacteristically cold winter and your ground freezes, you could lose some or all of your tubers. If you want to try overwintering but aren't sure if you'll be successful in your region, you could mulch the majority of your plants, then dig and store a couple clumps of each variety for insurance in case the winter is colder than expected.

# ANATOMY OF A TUBER

In order to grow and reproduce, a dahlia tuber must have three key components: the main body, which holds the nutrients and energy needed to produce next year's plant; an eye or eyes, which will eventually swell and sprout, becoming the stalks of the new plant; and a sturdy, unbroken neck that connects the two other parts. Without any one of these, your tuber won't grow. You might, for example, have a tuber with a healthy, visible eye and a good-size body, along with a broken neck. The neck will ultimately rot, so you should discard this tuber, since the sprout won't ever be able to access the nutrients from the body. Or you might have a large, firm tuber body and strong neck, but no eyes. This should also be tossed, because without an eye, it won't sprout.

In the early years I kept anything that looked like it might be viable. Eventually, after sorting through hundreds of rotten or eyeless tubers, I finally learned my lesson. Now I keep only tubers with all three components. It's better to be ruthless when sorting and save yourself storage space and time in the long run.

## SIZE DOESN'T MATTER

It's important to highlight a common misconception that the size of a dahlia tuber correlates to the ultimate size and strength of the plant. I have seen many new growers pass over small but healthy tubers for fear that they would not produce a strong plant, but this is an unnecessary worry. Every variety is different: some produce long, skinny tubers, while others produce tubers that are the size and shape of a potato. Even on a single clump of the same variety, you'll find varying sizes of tubers, and as long as each has a body, an eye, and a strong neck, it will produce abundantly.

58

EYES

NECK

BODY

1.

2.

3.

4.

One of the most rewarding parts of growing dahlias is digging up and dividing the clumps of tubers every autumn, and this is the most common way to multiply your stock. Even after so many years of growing, I'm still amazed that a single tuber that I planted in spring can produce 3 to 10 tubers just half a year later. Few flowers give you a greater return than dahlias. Whether you're dividing just a few clumps or an entire field, the process is the same. And while it can be a bit time consuming, you will be greatly rewarded for your efforts.

Dividing dahlias is messy, so work on a table in a location that doesn't have to remain tidy, stays above 40°F (4.4°C), and has ample lighting.

---

SUPPLIES  Waterproof gloves • Rain gear • Tuber clumps • Brass fitting with shutoff valve for your hose • Needle-nose snips (or heavy pruners, if preferred) • Permanent marker • Label or masking tape • Plant tag or wooden stake

1. Wearing waterproof gloves and rain gear, spray off a tuber clump with a strong stream of water, working from the inside out so the dirt falls away from the necks of the tubers. There's nothing worse than trying to find eyes in a muddy mess, so make sure to thoroughly clean each clump. I prefer to use a brass fitting with a shutoff valve on my hose rather than a traditional nozzle, because holding a trigger spray nozzle for a long period of time will make your hands cramp. The better you wash your clump, the easier it will be to divide later.

2. After washing all your clumps, set them in a frost-free place to drip dry. I typically wash clumps 1 to 2 days before I plan to divide and store them. You should try to wash your tuber clumps as close to dividing time as possible, because once they are clean, they will start to dry out within a few days and eventually shrivel, and may not be viable.

3. When clumps are clean and dry, use your pruners or snips to split clumps in half (you don't have to be delicate in doing this) to make them more manageable. It's easy to inadvertently mix up varieties when multiple people are working at the same table, so in order to keep all clumps of a variety together, work with a single variety at a time.

4. Once a clump is halved, you can split again so you have more manageable chunks to work with.

*continued* $\longrightarrow$

61

5.

6.

7.

8.

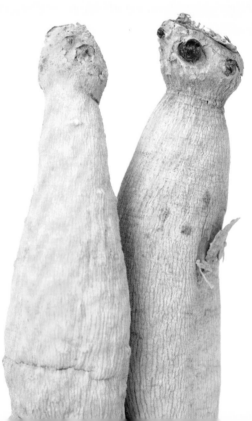

5. Throw away any tubers that have broken necks, rot, or major damage. This initial culling makes finding the good tubers a whole lot easier.

6. After removing broken and damaged tubers, start at the top of the clump and cut away viable tubers, using sharp snips to slice through the crown and separate individual tubers from the section.

7. Make sure that each tuber has a strong neck and visible eye. (This detail shows why needle-nose snips are so important: they allow you to cut precisely between the tubers and the crown of the clump, keeping the eyes safely attached to the tubers.) The eye is what will develop into a sprout.

8. Hone your skill at identifying the eyes on tubers. Eyes can be tricky to spot, especially if you're dividing right after digging, because plants will be more dormant. Some varieties have very obvious eyes, and others are nearly impossible to detect. These tubers are in varying stages of dormancy. The one on the left has eyes that are hard to see, while the tuber on the right has eyes that are starting to swell. The more you practice dividing, the better you'll get at identifying these. Remember that if a tuber doesn't have an eye, it will not grow.

*continued* $\longrightarrow$

63

9.

10.

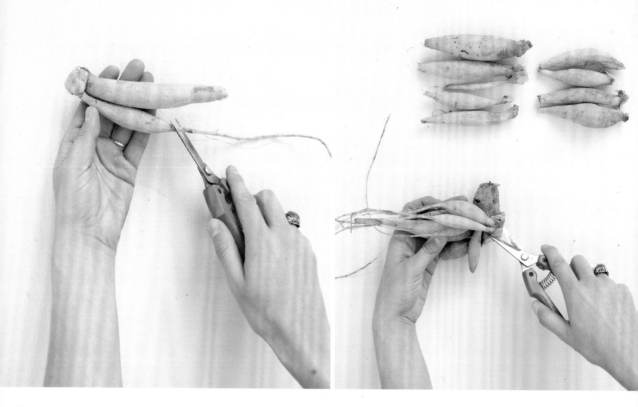

11.

12.

9. As you're dividing, clean up the tubers as you go. Trim off the roots and any jagged or excess chunks around the neck, as they are prone to rot and shriveling. Wherever you make a cut on the tuber, it will callus over.

10. Continue dividing and cleaning, keeping only the tubers that each have a body, eye, and strong neck, until you've divided the entire section.

11. After dividing all of the tubers of a particular variety, let them lie out and dry thoroughly before labeling and storing them. If tubers are stored while still wet, they will be prone to rot, and if they're left out to dry for more than a few days, they will start to dehydrate and shrivel up. I typically wait about 24 hours so that tubers are definitely dry but don't start to shrivel.

12. After all of the tubers are dry, label them. If you have large quantities of a particular variety, you can simply label the outside of the storage container, and add a labeled plant tag or stake in the container as a backup in case the outside label is damaged. But if you have smaller quantities, it's helpful to individually label the tubers with a permanent marker.

65

# STORAGE

Nailing down the perfect method and location for storing your dahlias in winter can be tricky. The most important factors to keep in mind when it comes to winter storage are temperature and humidity. Dahlias are extremely cold sensitive, and tubers need to be kept in a stable environment that has a constant temperature that's cold but doesn't freeze, ideally between 40°F and 50°F (4.4°C to 10°C). If the temperature drops too low, tubers will freeze and ultimately turn to mush, and if the temperature gets too warm (above 50°F [10°C]), tubers will think it's spring and start to sprout. I've found that the coldest corner of a basement, or an insulated garage that never gets down to freezing temperatures or has a space heater or other backup heat source during the coldest parts of winter, both work well.

Maintaining proper humidity is also crucial. If stored wet, tubers will mold and possibly rot; if they are stored too dry, such as uncovered in a cardboard box or paper bag, they will shrivel up and may die. A little shriveling is perfectly normal and won't affect the viability of the tubers. But if left unchecked, some tubers will shrivel so much that by the time you pull them out in the spring they will be past the point of revival. So the best way to prevent shriveling is to store tubers in a plastic container with some type of medium, as detailed ahead. If you find yourself with shriveled tubers, though, not all hope is lost: As long as they still have some moisture left in the body (which you can feel by gently squeezing them), an intact neck, and a visible eye, they will likely still grow. But if you're finding that a high percentage of your tubers are shriveled, you'll want to upgrade your storage process for next year.

There are numerous techniques for storing tubers, and I have tried them all. Each has its pros and cons, and after many years of experience, I have narrowed them down to my two favorite ways: storing in plastic wrap or in bins, which are both detailed ahead. Whatever method you choose, it's important to inspect your tubers monthly over the winter and dispose of any rotten or moldy tubers so that they don't contaminate the rest of your stock.

1.

2.

3.

4.

# HOW TO STORE DAHLIAS IN BINS

For keeping large quantities of tubers, and to minimize waste, storing tubers in containers with a naturally derived medium is my go-to choice. I have experimented with many different media including peat moss, sand, wood shavings, straw, and vermiculite, and I have found that coarse vermiculite, while more expensive, is the easiest to use and most effectively maintains humidity. It's important to note that vermiculite can be dusty, so if you're working with it a lot, you should wear a dust mask.

The number and size of a given variety of tubers will determine the size of storage container you need. Know that paper bags and cardboard boxes are highly breathable and will allow too much moisture to escape, so I strongly recommend storing in plastic containers or bags. Ten or fewer tubers will typically fit in a gallon-size resealable bag, whereas 12 to 25 tubers need a shoebox-size plastic bin. To ensure that excess moisture can escape during storage, I like to drill 2 to 3 holes in the upper portion of each bin. If, when using resealable bags, you notice water droplets forming on the inside, be sure to vent the bag.

1. Fill a plastic bin with a 1 to 2 in (2.5 to 5 cm) layer of vermiculite making sure to completely cover the bottom.

2. Set your tubers on top of the vermiculite so that they are close but not touching. If the tubers touch and one molds or rots, it will likely spread to the other tubers.

3. After you have a full layer of tubers in place, cover them with another 2 in (5 cm) layer of vermiculite, making sure that all the tubers are completely covered.

4. Continue adding layers of tubers and vermiculite until the bin is full, ending with a layer of vermiculite over the top. I equate the process to making lasagna. Be sure to tuck a plant tag or label on top of the vermiculite inside the bin (in case the outside label is damaged) and label the bin exterior with masking tape. Cover the bin with its lid. When you pull out your tubers in spring, brush off the vermiculite and save it for later use.

SUPPLIES   Plastic bins with lids • Vermiculite or other media • Plant tags or labels • Masking tape or labels for bin exterior • Permanent marker

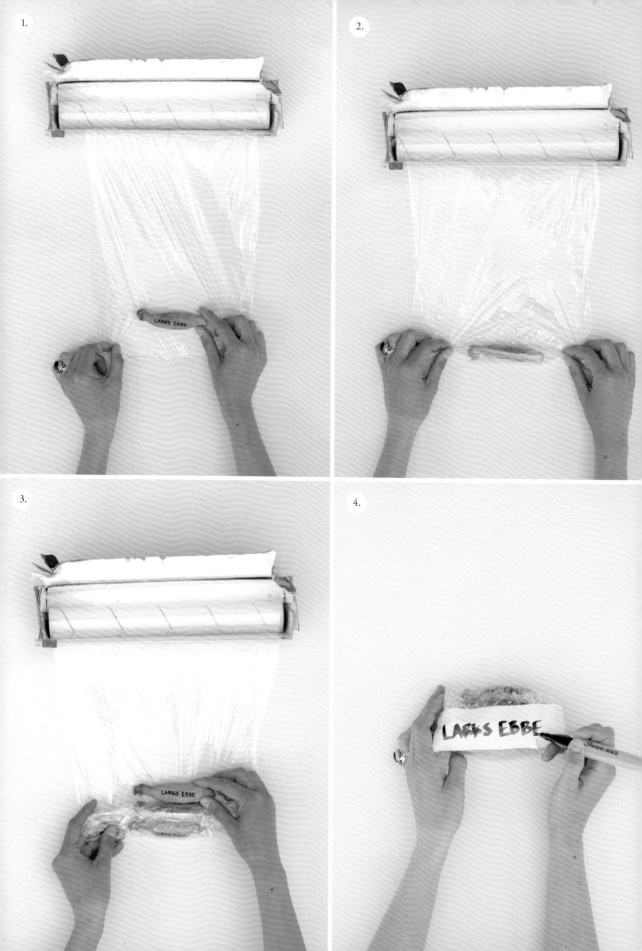

# HOW TO STORE DAHLIAS IN PLASTIC WRAP

A few years back while attending our local dahlia club meeting, I heard about a new ultra-easy storage method—wrapping tubers in plastic wrap—that many of the members were converting to because of its high success rate. This is an ideal technique if you have only a small quantity of tubers to keep or are short on space, as it doesn't require so many bins and bulky bags filled with media. Instead, you're left with tidy little bundles of tubers, each fully surrounded by a thin layer of plastic. The plastic helps hold in moisture and also prevents the spread of rot if a tuber in the bundle goes bad. Nearly every grower that I've talked to who uses this method finds that 90 percent or more of their tubers are viable after being stored this way. The downsides to this technique are that you have to spend time wrapping and unwrapping the tubers, and you create a fair amount of waste in the process.

SUPPLIES  Plastic wrap • Masking tape or sticky label • Permanent marker

1. On an 18 in (46 cm) section of plastic wrap, still attached to the roll, place the first tuber 3 to 4 in (8 to 10 cm) from the cut end.

2. Fold the cut end of the plastic wrap over the tuber and press it against itself. Roll the tuber toward the uncut end a few times, enclosing the tuber in a few layers of plastic.

3. After the first tuber is sealed, add additional tubers one at a time, rolling as you go so that each tuber is enclosed in a layer of plastic wrap and is not touching the adjacent tuber. Continue pulling out more plastic wrap as needed.

4. When you've securely wrapped a manageable bundle of tubers (I normally bundle in quantities of 10), fold in the sides of the plastic wrap so that the tubers don't fall out. Your finished bundle should resemble a burrito. Label the bundle with the variety name even if your tubers are labeled individually, as this will help you locate varieties faster. Check monthly by examining the wrapped bundles for moldy or mushy tubers. Toss any tubers that are affected.

71

CHAPTER FOUR

# ADVANCED TECHNIQUES

Once you become comfortable with growing dahlias from tubers, you may want to try your hand at a few more advanced methods. While the most common way to propagate dahlias is by dividing tuber clumps, as mentioned previously, taking cuttings is another great but lesser known method to multiply your stock.

A few years ago, I saw a photo of an incredible soft pink dahlia called 'Castle Drive' on Instagram, and I went on a mad hunt to find tubers for it. For months, I scoured all of the North American sources I could find, without any luck. Over 100 calls, emails, and dead-end searches later, I finally gave up hope of ever finding them. Then one day in late April 2016, out of the blue, I found a box with 20 perfect tubers waiting in my mailbox, with no return address. To this day I still don't know who sent them to me! I've taken very, very good care of that original stock, and was able to multiply those original tubers from 20 to over 2,000 in just two growing seasons through the process of taking cuttings. In addition to the incredible soft pink/blush coloring, they are one of the first varieties to bloom, and the flowers are long lasting with strong, tall stems. If I had to pick a favorite variety (we have nearly 800), this beauty would surely be it!

# PROPAGATION METHODS

### GROWING FROM CUTTINGS

One of the fastest ways to increase your stock of coveted or expensive varieties is by potting up and bringing tubers into growth early, then taking cuttings as sprouts emerge. Plants grown from cuttings will be the same variety as the tuber they're cut from. Unlike dividing tubers, which requires an entire growing season for tubers to multiply underground, you can produce a large, healthy rooted cutting that's ready for planting in about 6 to 8 weeks.

Cuttings do require some special equipment and dedicated time, but the return on your investment can be tenfold, because a single tuber can produce 10 to 20 cuttings in a few months. In addition, you can plant any tubers that you take cuttings from; just be sure to bury the tubers like you would any others, after all danger of frost has passed in spring.

Cutting-grown plants generally come into flower 2 to 4 weeks earlier than those grown from

tubers. You can count on them to develop at least 2 to 3 viable tubers by autumn, which isn't quite as abundant as tuber-grown plants but still increases your stock significantly.

## GROWING FROM SEED

While dahlias are most often grown from tubers so that farmers and gardeners get the exact varieties they want, you can also grow them from seed quite easily. Many sources sell dahlia seeds in mixes that are generally grouped by type such as pompon, small flowered, and collarette. In addition, you can save your own seed at the end of the growing season (see How to Harvest Seed, page 101). The main thing to understand about growing from seed is that you'll get a mix of dahlias in different shapes, colors, and sizes rather than a specific variety. These mixes are an inexpensive way to get your hands on a lot of dahlia plants fast, but keep in mind that you never know what you're going to get.

# HOW TO TAKE CUTTINGS

When taking cuttings, keep in mind that they are very tender and require a warm, bright space to thrive, so unless you have a heated greenhouse, you'll need to construct a growing station. A table or shelf with a heat mat on top and some inexpensive shop lights (LED or fluorescent) hanging from above will do the trick. I converted a corner of my heated garage into a cutting station and produce thousands of cuttings every winter. Once you get the hang of it, you'll be hooked and have more dahlias than you know what to do with.

SUPPLIES   Coarse potting soil • Pots that are at least 3½ in (9 cm) wide and 5 in (13 cm) tall • Bottom trays with drainage holes • Plant labels • X-Acto knife • 72-cell seed tray filled with coarse potting soil • Pencil, chopstick, or bamboo skewer • Rooting hormone (I prefer gel types) • Flat-bottomed tray with no holes • Clear dome lid • Heat mat • Shop lights (LED or fluorescent) • Small pots

1. In late winter, pull out the tubers you want to multiply and plant them vertically in soil-filled pots that are at least 5 in (13 cm) tall, leaving at least 1 in (2.5 cm) of the neck poking out of the soil so that the eyes are easily accessible. Label each pot with the variety name and date potted up, and set them on a plant tray with drainage holes.

2. Place the potted tubers in a 65°F to 70°F (18°C to 21°C) spot to wake up; I keep ours in a heated hoop house. Note that your tubers should not be placed on a heat mat at this point, since that would encourage rotting from below. The first year I took cuttings, I lost almost half of my stock to this mistake. The process of waking up the tubers usually takes 2 to 3 weeks. Once the eyes swell and sprouts begin to emerge, they should be moved to an area that stays warm and has plenty of light, such as a heated greenhouse, or indoors under shop lights.

3. Once the sprouts are 3 to 4 in (8 to 10 cm) tall, you can start taking cuttings. (If sprouts are shorter or much taller than this, their chances of rooting will be greatly diminished. I regularly monitor the tubers to catch the sprouts at their optimal size.) As cleanly as you possibly can, use an X-Acto knife to gently slice off the sprout where it connects with the tuber.

4. It's important that you make your cut flush with the tuber so that you don't cut out the growing eye or cut too high into the stem of the sprout where it is hollow. If the cut was made at the proper location, you'll see a white ring in the center of both the sprout and where it was removed.

79

*continued* $\longrightarrow$

5. Once you have collected a small pile of cuttings, carefully remove the lower two to three sets of leaves from each so that you have at least 1 to 2 in (2.5 to 5 cm) of cleaned stem to work with. If the leaves that sit below the soil line are not removed, they will rot. Do not let your cuttings sit out for more than 15 to 20 minutes, or in direct sunlight, because they will wilt and likely won't recover.

6. Fill a seed tray with coarse potting soil, water it deeply, and poke holes in the center of each cell, using a pencil, chopstick, or bamboo skewer. Make sure your holes go all the way to the bottom of the tray. You will insert your cuttings into these holes, and poking them ahead of time will help ensure that the tender cuttings aren't damaged when you place them.

7. Dip the bottom 1 in (2.5 cm) of the cleaned cuttings into some type of rooting hormone. There are many products available, but I find that gel is easiest to use and the least messy to work with, though it smells terrible!

8. Insert dipped cuttings into the pre-poked holes until they touch the bottom of the tray. This will ensure that they have contact with the heat mat and root more quickly. Using your fingers, press the soil firmly around the cuttings so that there are no gaps between the soil and the stems. Continue until the tray is full, taking care to label individual varieties as you go.

*continued* ⟶

9. Add about 1 in (2.5 cm) of water to your no-hole tray and place the tray of cuttings inside. Set the dome lid over the top of the tray; if you prefer not to use a dome lid, you'll need to mist the cuttings with water (a small spray bottle works great) 3 to 4 times a day so they stay consistently moist. Put the tray on a heat mat set at 70°F (21°C), under lights. Suspend the lights 1 to 2 in (2.5 to 5 cm) above the top of the dome lid, and leave them on for 14 to 16 hours per day (you can use a timer if you like). Check cuttings daily and remove any that turn yellow or moldy. Make sure to keep the water level in the tray at 1 in (2.5 cm) deep. If you have too much water, the cuttings will rot; if you have too little, the cuttings will wilt.

10. I find that, given the ideal environment, it takes 12 to 14 days for cuttings to develop white roots, which indicate that they're growing. After years of propagating tens of thousands of plants, I have found that the sprouts will take on a grayish cast, and a whole tray will look as if it's starting to go downhill, about 2 days before they send out their first white roots. It's like clockwork: every time I think a tray of cuttings is starting to die, they inevitably send out roots 1 to 2 days later. You can gently pull up the cuttings to check on root development.

11. Once the cuttings have developed enough roots to keep their soil intact (1 to 2 weeks after they've developed white roots), move them into larger pots. For each cutting, fill a small pot with coarse potting soil and make a hole in the center (I find that a butter knife works well for this.) Slip the rooted cutting into the hole and press the soil firmly around the baby plant. After potting up the cuttings, water well, and place them either back under the lights or in a heated greenhouse. It's important that cuttings are held in a warm, bright environment (over 60°F [15.6°C]) so that they can continue to actively grow.

12. After 3 to 4 weeks, the rooted cuttings should have filled out their pots and are ready to plant into the garden. Be sure to wait until all threat of frost has passed before planting outside because cuttings are very tender—in Washington, this is around Mother's Day. If it is still too cold to plant outdoors but your cuttings are becoming rootbound, you can transplant them into larger pots while you wait for the weather to warm.

To learn how to plant rooted cuttings in the garden, see Planting, page 34.

In general, you'll want to start your dahlia seeds in early to mid-spring, about a month before the last spring frost. Like cuttings, seeds need a relatively warm, bright space, such as a greenhouse, to germinate and develop. If you don't have a greenhouse, you can create a similar environment indoors as detailed ahead.

---

SUPPLIES  72- or 128-cell seed starting tray • Coarse potting soil • Pencil, chopstick, or bamboo stake (optional) • Dahlia seeds • Vermiculite • Watering can or hose with water wand • Bottom tray with drainage holes • Plant labels • Clear dome lid • Heat mat • Lights (LED or fluorescent)

1. Sow seeds about 4 to 6 weeks before your last spring frost. If sown earlier than this, plants will become too large and won't transition well. Fill a cell tray with premoistened coarse potting soil, and tap the tray on a table to remove any air pockets. Fill in any of the cells that need a little extra soil.

2. Use the tip of your finger, a pencil, a chopstick, or a bamboo stake to make a ¼ in (0.6 cm) indent in the middle of each cell of soil.

3. Place 1 seed per cell in the indentations you made, and continue until all the seed has been sown or your tray is full.

4. Cover the seeds with a fine layer of vermiculite or potting soil, just enough to cover them. Don't bury them too deeply.

85

*continued*  ⟶

5. Once your seeds have been sown and covered, gently water them in with a water wand or watering can. Place your seed tray in a bottom tray with drainage holes.

6. Label your tray with the name of the mix or of the parent variety (if you harvested the seeds or the info is available from your seed source). Then place the dome lid over the top to maintain heat and humidity. Set the tray on a heat mat set at 70°F (21°C) under lights, or in a greenhouse. If under lights, suspend the lights 1 to 2 in (2.5 to 5 cm) above the top of the dome lid, and leave them on for 14 to 16 hours per day (you can use a timer if you like).

7. Once the seedlings sprout, remove the dome lid and take the trays off the heat mat. Continue to grow the seedlings in a warm, bright environment until it's safe to plant them outside. Please note that a windowsill does not provide adequate light—plants there will become leggy and weak and won't transition well into the garden.

8. Once the weather has warmed and all danger of frost has passed, plant the seedlings in the garden. Slugs and snails love the tender young plants, so preventatively combat them by sprinkling slug bait around your seedlings. I use Sluggo, an organic choice that is kid- and pet-safe.

# HYBRIDIZING

Dahlia breeders are passionate, competitive people who spend countless hours in their gardens trying to create the next big thing. Because of their patience and commitment, the world has been blessed with thousands of incredible varieties to grow and enjoy. Historically, most dahlia hybridizing has been done in the backyards of hobby gardeners who have focused their efforts on creating varieties that will win awards at society shows. While the show world is fascinating in its own right, I've noticed that many of the award-winning dahlias do not necessarily fit the current style and what's in fashion with floral designers, flower farmers, and home gardeners these days.

After more than a decade of growing cut dahlias professionally, and feeling constantly disappointed by the lack of sought-after colors (such as blush, champagne, smoky peach, and muddy raspberry), I decided to try my hand at hybridizing in hopes of creating more varieties to fill this need. While the breeding world is typically quite secretive, because hybridizers understandably want to win competitions, I have been blessed with two generous mentors who have been willing to share their time and experience with me. Ken Greenway, one of the most prolific dahlia breeders of our time, lives nearby and has so generously welcomed me to his garden on numerous occasions to show me firsthand how he creates all his big show winners with such a high rate of success. Likewise, Kristine Albrecht of Santa Cruz Dahlias (see A New Breed, page 97) has provided a wealth of information and has made the world of breeding less mysterious and much more attainable.

At its simplest, hybridizing involves collecting seed from desirable plants (that you have either hand-pollinated or left for the bees to pollinate) and planting the seeds to see what new varieties emerge. No two seeds will produce the same results, even if they came from the same seedpod. Every seed will produce a different plant. When it comes to dahlias, the only way to get an exact replica of the original plant is by cloning it from either a tuber or a cutting. Seeds, on the other hand, will produce brand-new varieties; each may possess some of the original traits of its parents but will ultimately be an entirely new creation.

Hybridizing methods vary in complexity, depending on how serious you want to get. While most dahlia enthusiasts have divided their own tubers and taken cuttings, very few have experimented with saving their own seed and breeding new varieties. As gardeners, we are so well trained to harvest and deadhead our flowering plants to increase our blooms that we rarely let them go to seed. But by leaving a few dahlia blooms to mature on the plant and develop seeds, you'll have the opportunity to give hybridizing a try.

It's important to know that no matter how much time and energy you pour into hybridizing, it's still kind of a crapshoot in the end. Most breeders say that, despite their best efforts, they rarely keep more than 1 percent of their trial seedlings, and then just a fraction of those in the longer term. I've found this to be true for myself as well—I expect to save at most 1 promising seedling from every 100 we trial. And while those odds can seem quite dismal, the possibility of breeding the next 'Café au Lait'—that sought-after, creamy-blush dinner plate variety—is well worth the effort.

Please note that what follows is very basic information and intended as an overview. If you want to go down the hand-pollinating path or dig deep into hybridizing, I recommend joining a local dahlia society to learn more.

## CHOOSING PARENT PLANTS

You can collect seeds from any of your dahlias, but you'll have a greater chance of success by harvesting from those that have well-defined form, a color you like, a vigorous growth habit, and overall good plant health.

To increase your chances of creating a new treasured variety, starting with the best parent plants will give you the best results, but figuring out which parent plants to use can be daunting. Not every dahlia variety makes ample seed. For example, 'Sherwood's Peach' is one of the most beautiful smoky-peach dinner plate dahlias, but its seedpods have a tendency to rot before

maturing. On the other hand, 'Café au Lait' produces a bumper crop of big fat seedpods despite its large showy blooms. So you could use 'Café au Lait' as the seed parent and plant it next to 'Sherwood's Peach' in hopes of imparting its incredible coloring to their offspring.

Fortunately, there are some tried-and-true varieties that many hybridizers look to because they reliably produce a bumper crop of seed. By growing varieties that consistently set seed and crossing them with other varieties that possess the traits that you would really like to see in future generations, your odds of creating a treasured new variety will be greatly improved.

The following varieties are known to produce an abundance of seed, and while they are not necessarily favorites that we've included in our Variety Finder (page 137), all are great candidates for breeding.

**Formal Decorative**

'Clearview David'
'Peaches N' Cream'
'RJR'
'Tall Firs Keltie'

**Informal Decorative**

'Café au Lait'
'Hollyhill Black Beauty'
'KA's Cloud'
'KA's Khaleesi'
'Kidd's Climax'
'Walter Hardisty'
'Wyn's Moonlight Sonata'

**Ball, Mini Ball, Pompon**

'Chimacum Davi'
'Chimacum Troy'
'Cornel'
'Greendor'
'Irish Glow'
'Jomanda'
'Mary's Jomanda'
'Ms Kennedy'
'Narrows Kirsten'
'Odyssey'
'Suncrest'

**Anemone**

'Dad's Favorite'
'Daisy Mae'
'Eileen C'
'Fancy Pants'
'Mexico'

**Collarette**

'Appleblossom'
'April Heather'
'Ferncliff Dolly'

**Stellar**

'AC Rooster'
'Alloway Candy'
'Camano Pet'
'Irish Pinwheel'

**Cactus**

'AC Conan'
'AC Cougar'
'AC Steve'
'Badger Twinkle'
'Kenora Challenger'
'Weston Spanish Dancer'

Breeding Patch Diagram

CAFÉ AU LAIT    SHERWOOD'S PEACH    CAFÉ AU LAIT    KA'S CLOUD    CAFÉ AU LAIT    SHERWOOD'S PEACH    CAFÉ AU LAIT

## PLANNING YOUR BREEDING PATCH

One of the most important things to keep in mind when hybridizing is to keep like forms together in the garden and provide enough separation between forms to prevent crossing. If you don't keep the different forms isolated from each other, the offspring will be all over the board. For example, by planting rounded forms (balls, mini balls, and pompons) near each other, most of the offspring from those crosses will retain the original rounded form. But if you plant some open-centered varieties in close proximity and the bees get to them both, you'll ultimately get a crazy mix of rounded flowers with single open centers.

In order to keep different forms from crossing with each other, you'll need to separate them by at least 50 ft (15 m). I know some clever hybridizers on small lots who are able to grow two separate breeding patches by planting one group of forms in the front yard and another group of forms in the backyard. As long as their neighbors don't grow dahlias, this plan works. If you have a really small space you can concentrate on hybridizing one form at a time or give hand-pollinating a try.

When designing a breeding patch, it's important to note that bumblebees, which are the best dahlia pollinators, have a tendency to fly in a straight line down a row, stopping at one flower and then the next, working their way steadily through the row. So rather than planting long straight rows of individual varieties, planting in smaller blocks is preferred. For example, if I have 16 plants of 'Café au Lait' that I want to use in my breeding patch because they are such good seed producers, rather than plant all 16 in a long straight row, it's better to divide the total up and plant them in smaller blocks, alternating them with other varieties in between. So in a row of informal decoratives I might plant

4 'Café au Lait' and then 4 'Sherwood's Peach', 4 more 'Café au Lait', 4 'KA's Cloud', 4 'Café au Lait' and then 4 'Sherwood's Peach', ending with another 4 'Café au Lait' (see Breeding Patch Diagram, page 92). That way, as the bees fly down the row, they are gathering pollen from each small grouping of plants and transferring it to the next, increasing cross-pollination throughout.

## HAND-POLLINATING

After a few successful seasons of hybridizing with the help of bees, some of the most diehard breeders decide to take it to the next level and start making variety crosses by hand. The process of hand-pollination is quite labor intensive and time consuming, so I wouldn't recommend this method to beginners, but it can increase the odds of creating a more desired outcome.

To hand-pollinate, isolate both of the varieties that you would like to cross so the bees can't access them. To do this, cover the flower buds with large organza bags and securely tie them at the base to prevent insects from accessing the flowers. Keep in mind that the larger the flower, the bigger the bag needs to be. After 1 to 2 weeks, the covered blooms will fully ripen and start producing pollen. Once the pollen centers are present, remove the organza bag from the bloom, taking care to not let the bees find it while you're working, and collect pollen with a small paintbrush and set it aside. Then cover the bloom again. Take the pollen-covered paintbrush to the flower you wish to cross, gently uncover the bloom, and transfer the pollen from the brush to the center of the flower. Then cover that bloom with the organza bag again. Be sure to wash the paintbrush between pollination sessions to avoid cross-contamination. For the best chance of success, repeat this process daily for at least 3 to 4 consecutive days. If your efforts

93

are successful, you will be able to harvest seed-pods from the covered blooms in 4 to 6 weeks. It's important to note that hand-pollinated pods tend to produce a lot less seed than those visited by the bees.

## HARVESTING SEED

As blooms ripen and fully open, you'll notice that even the tightest ball varieties will eventually expose their open centers, which are loaded with golden pollen. As individual blooms are pollinated, you'll see their outer petals start to wither. Some varieties naturally shed their petals easily, while others, including many of the larger dinner plate varieties, need a little extra help so that the older petals don't stick and rot the seed-pods. I regularly go through the breeding patch and gently remove the older, fading petals from the backside of the flowers so that the seedpods can fully ripen without risk of rot. It can be easy to get carried away and want to strip away all of the petals at once, but it's important to take great care and only remove those petals that easily give way when gently coaxed off. After pollination occurs and the petals fall away, you'll notice that the pollen center will eventually turn brown, and over time the seedpod will slowly close up around it. Once this happens, the seedpod will become pointed and slowly change colors, from green to gold, and, if left on the plant for long enough, eventually turn brown and papery. Seedpods can be harvested at different stages of ripeness depending on your climate and how late in the season it is. Learn how to harvest seed on page 101.

## PLANTING AND EVALUATING

After a season's worth of work, the fun really begins, and you will finally be able to see the fruits of your labor. In mid-spring, pull out all of the seeds that you've saved from the previous season and decide how many you'd like to plant. For sowing instructions, see page 85.

Keep in mind that you can fit a lot more seedlings into a planting bed than tuber-grown varieties because they don't require as much space. While it's fine to give your seedlings plenty of room to spread out, hybridizers generally space their breeding seedlings just 6 in (15 cm) apart down a row and squeeze 4 or 5 rows into a planting bed, so a bed that's 3 ft (91 cm) wide and 10 ft (3 m) long can hold up to 100 seedlings. The reason for the tight spacing is that, in year one, all you're looking for is whether or not the plant makes a beautiful flower, so you don't need to give up valuable garden space for this purpose. You really only need a single flower to determine if the tuber is worth saving. This past season we grew 13,000 seedlings and ended up keeping only about 500. While every flower in the seedling field was beautiful, there were a very small number of blooms that had the coloring I was after.

Once seedlings are blooming in earnest, I comb the patch every week looking for new treasures to emerge and flag those that have potential. I've found that in a giant sea of flowers, it can be tricky to keep track of which varieties I've already flagged, so I place a 6 ft (1.8 m) tall bamboo stake next to the chosen variety, pushing it down into the ground and tying the plant to the stake with fluorescent flagging tape. I then tie another piece of flagging tape around the base of the plant at ground level so at digging time we can easily find them. My weekly comb through the seedling field feels like a treasure hunt every time.

At the end of the season after the first autumn frost, I go through the seedling field and dig up only the varieties that have been flagged. You'll notice that the tuber clumps are much smaller than normal dahlias and only produce 1 to 3 tubers per clump from a first-year seedling. I normally divide these clumps, keeping the varieties separate from each other, and give each one a number instead of a name. Many growers don't even bother dividing them and just replant the miniature clump back out in the garden the next season for observation.

The next season is when things get really interesting, as you can see which varieties rise to the top and which pale in comparison. I've found that many of the varieties I end up saving aren't nearly as wonderful as I thought when being evaluated the second year. When it comes to hybridizing, it's important to be a little bit ruthless and only keep the very best performers—varieties that have beautiful coloring, the flower form you're after, strong plant health, and good tuber production.

If a variety performs well and fits all of your criteria in the second year, it then gets to move to the next level. By the third year, you'll have a pretty good idea if a variety is a true winner or not and can start dreaming up potential names. From here, it's up to you whether or not you want to enter your creation into the show circuit or if you want to introduce it to the world without officially registering it.

My first year dabbling in hybridizing I planted 1,600 seedlings. From that patch, about 150 were flagged as showing potential. And the following year, of those, only 27 made the cut. After one more season of growing out, I settled on merely a dozen standouts. The project spanned 4 years in total, and while none of my creations were the next 'Café au Lait', the project was a total blast, and I've learned so much in the process. While hybridizing is a long-term commitment, if you've got a little extra space and some time to devote to breeding, it can be the ultimate treasure hunt.

## A NEW BREED

On a ¼ acre (1,012 square m) plot in suburban Santa Cruz, California, where she grows more than a thousand dahlias, hybridizer Kristine Albrecht is intent on creating varieties with the coloring that floral designers are after.

The owner of Santa Cruz Dahlias, president of the Monterey Bay Dahlia Society, and executive board member of the American Dahlia Society (ADS), Kristine has had a number of dahlia introductions that have taken top awards at ADS and other shows. She explains that the vast majority of serious hybridizers grow between 200 and 4,000 dahlias from seed each year. Some sell their hybrids but many aren't focused on selling tubers or cut flowers to consumers. Instead, they aim to breed varieties with traits—such as perfect color or form as governed by ADS guidelines—that will win at dahlia shows. "This approach excludes the complex colors and unusual forms that work so well as cut flowers and for designers," Kristine says.

Dahlias in muted shades such as soft peach, pale pink, and variations of white, or that have petals with imperfect form or color patterning, won't make the ADS cut, even if they're ones floral designers love. An example is the beloved 'Cafe au Lait'. "It's one of the most popular dahlias in the world, but it's not considered a top show flower by ADS," Kristine explains.

Kristine's willingness to pass along knowledge has gotten a lot of budding hybridizers pointed in the right direction. And fortunately, as she and other breeders discover new "designer" varieties—even if they aren't destined for the show bench—they may soon be available to bloom in your own garden.

1.

2.

3.

4.

# HOW TO HARVEST SEED

In our cool climate, it takes many weeks for dahlias to set seed, and I've found it best to stop harvesting flowers from my breeding plants at least 8 weeks before the first autumn frost so as to not interfere with their pollen and seed production, and I'm guaranteed lots of ripe, mature pods.

SUPPLIES   Pruners • Jar of water • Seed tray, shallow cardboard box, or paper plate • Paper envelopes • Pen

1. Keep an eye on your dahlias to determine when they're ready for harvest. Once a dahlia has dropped its petals, its seedpod will begin to form and change color, from bright green to a warm gold and eventually light tan. For dark foliaged varieties, the pods will turn a chocolate-brown color.

2. Squeeze the seedpod to assess ripeness. If it drips, it is still too young.

3. Harvest seedpods. You can snap them off the stem, though I prefer to cut them with the stem still attached because it's easier to hold them while harvesting and you can also label the entire bunch with the variety name if you want.

4. If you harvest pods before they are fully ripe, place them with their stems still attached in a jar of water in a warm, bright place to finish maturing inside, which usually takes 1 to 2 weeks. After they are removed from the water, they should be laid out to dry completely.

*continued*  ⟶

5. Lay the freshly harvested pods out to dry. Set them on seed trays or shallow cardboard boxes in a warm, dry place. If you have a small amount of seedpods, you can also use paper plates.

6. Once the seedpods are fully dry and papery to the touch, they can be shelled by rubbing the pods between your fingers.

7. Separate the dark gray and black seeds from the papery husks and discard the excess debris.

8. Store seeds in paper envelopes in a dry place until it's time to sow them in the spring. Be sure to label the variety you collected seed from so you can trace any new treasures back to their original parents. (If you hand-pollinate and then cover the blooms, you'll know both parents; if bees do the pollinating, you'll only know the seed parent, not the pollen parent.)

103

# CHAPTER FIVE

# DESIGNING WITH DAHLIAS

While dahlias are a wonderful addition to the garden, the best part of growing them is being able to harvest an abundance of their stunning blooms and bringing them indoors to enjoy in arrangements. Few other flowers are as productive, rewarding you with armloads of blooms from midsummer through the first autumn frost. In Washington, we're able to harvest dahlias from the garden for nearly 3 full months, and almost every bouquet I make in that time frame features these gems.

When arranging dahlias, there are a few things to keep in mind. Dahlias make an excellent cut flower, but they are not quite as long lasting as other summer blooms, typically persisting for up to 5 days if picked at the right stage and if they've been allowed to rest in water for a few hours after harvest (see Harvesting, page 42). When you're combining them with other flowers and foliage, just know that they will be one of the first ingredients to fade. Large-flowered dinner plate varieties, like the revered 'Café au Lait', require a little extra care, since the petals bruise easily and the heavy blooms can break off the stem if handled too roughly. Dahlias are such showy blooms and look marvelous on their own, either arranged en masse or displayed individually, so one can appreciate their unique traits and features up close. In addition to holding their own in an arrangement, they play well with others and can be incorporated into any mixed bouquet. What's not to love?

Note that not all of the dahlias mentioned in the bouquets ahead are in the Variety Finder on page 137 because we simply couldn't squeeze every single dahlia we love into the space we had for this book! But know that all of the dahlias shown are ones I definitely recommend adding to your garden and bouquets.

# ALL HAIL THE QUEEN

Of all the dahlias I've grown, 'Café au Lait' by far is the most notable. She is lovingly referred to as the queen of dahlias, and it's easy to see why she has such a loyal and wide-reaching fan club. Many of the dinner plate–size dahlias can be fragile and a little difficult to work with. 'Café au Lait', however, if pinched early on, produces long, strong stems that hold up really well in large arrangements.

Working with dinner plate varieties can be intimidating because of their massive size. I find the best way to approach them is to go all in and embrace their over-the-top nature. For this arrangement, I limited the number of ingredients and let the 'Café au Lait' blooms take center stage. I filled a large Vermont sap bucket with three different kinds of antiqued hydrangeas to create a pillowy nest for the Cafés. Next I threaded in some large branching stems of pokeweed to add texture and scale to the display. The pink stems on the berried branches and the blush flowery sprays helped carry color throughout.

After the base of the arrangement was well established, I nestled the 'Café au Lait' blooms in the hydrangea pillow, making sure to twist and turn the flower heads so that they weren't all facing forward, because I think the backs and sides of the flowers are just as beautiful. To help break up the domed shape of this bouquet, I threaded in long, wiry stems of 'Honka Fragile' dahlias and used their star-shaped blooms to contrast the softness of the other fluffy ingredients. While picking the ingredients for this arrangement, I noticed a beautiful common weed, wild sorrel, that had great texture and the perfect hint of blush to complement the other ingredients, so I added that as well.

---

INGREDIENTS   Dahlia 'Café au Lait' • Dahlia 'Honka Fragile' • Hydrangea Bobo • Hydrangea 'Limelight' • Hydrangea Quick Fire • Pokeweed • Wild sorrel (*Rumex acetosella*)

POKEWEED

DAHLIA 'CAFÉ AU LAIT'

HYDRANGEA BOBO

DAHLIA 'HONKA FRAGILE'

WILD SORREL
(*RUMEX ACETOSELLA*)

HYDRANGEA
QUICK FIRE

HYDRANGEA
'LIMELIGHT'

# FIELDS OF GOLD

For many years, I have been on a quest to collect all of the most beautiful yellow-toned dahlias that I can get my hands on. Yellow can be tricky because most of the flower varieties available in this color are too bright or bold to mix with anything else. To date, I've grown more than 200 yellow dahlia varieties, and I've pared that back to keep in my collection about 40 that are the most versatile for arranging.

Through the process of writing this book, we organized nearly 800 varieties into 11 main color classes, which was more challenging than we ever thought possible. The yellow dahlias were the most surprising and difficult of all to categorize. After defining what we considered yellow, we went on to create subclasses including honey (a warm, peachy yellow), goldenrod (a cheerful, glowing yellow), canary (a bright, sunny, clear yellow), buttercream (which reminds us of fresh butter), highlighter (a fluorescent yellow with green undertones), and yellow with cranberry dusting (where flowers have darker centers and tips).

To embrace the full spectrum within this diverse color, I chose to display individual dahlia varieties in my favorite Farmhouse Pottery vase collection and arrange them vignette style—one of the fastest and easiest ways to create an impactful flower display.

INGREDIENTS  Dahlia 'Blah Blah Blah' • Dahlia 'Bloomquist Butter Cream' • Dahlia 'Blyton Softer Gleam' • Dahlia 'Born Sty' • Dahlia 'Bracken Sarah' • Dahlia 'Buttercup' • Dahlia 'Citron du Cap' • Dahlia 'Clearview Citron' • Dahlia 'Golden Scepter' • Dahlia 'Hamari Gold' • Dahlia 'Karmel Korn' • Dahlia 'Lakeview Peach Fuzz' • Dahlia 'Lucky Ducky' • Dahlia 'Sandia Sunbonnet' • Dahlia 'Skipley Moonglow' • Dahlia 'Westerton Lillian' • Unnamed seedling

DAHLIA
'GOLDEN SCEPTER'

DAHLIA 'BLAH
BLAH BLAH'

DAHLIA
'CITRON DU CAP'

DAHLIA 'SKIPLEY
MOONGLOW'

DAHLIA
'KARMEL KORN'

DAHLIA 'SANDIA
SUNBONNET'

DAHLIA 'HAMARI GOLD'

DAHLIA 'BRACKEN
SARAH'

DAHLIA 'WESTERTON
LILLIAN'

DAHLIA 'LAKEVIEW
PEACH FUZZ'

DAHLIA
'CLEARVIEW
CITRON'

DAHLIA
'BLOOMQUIST
BUTTER CREAM'

DAHLIA 'LUCKY
DUCKY'

DAHLIA
'BORN STY'

DAHLIA 'BLYTON
SOFTER GLEAM'

DAHLIA
'BUTTERCUP'

UNNAMED SEEDLING

# PEPPERMINT SWIRL

I've always had a soft spot for red-and-white variegated dahlias, but could never figure out how to incorporate them into arrangements without them sticking out like a sore thumb. It wasn't until seeing a bouquet a friend received as a gift, which featured flowers in shades of red and white, that it finally occurred to me: the best way to display these fun novelties is all on their own.

When you have flowers that you want to highlight but that don't necessarily play well with others, showing them off en masse is a great solution. This arrangement came together in less than 10 minutes. I intentionally chose a vase with a slightly tapered neck to help keep the flowers more upright, since I didn't use any foliage or filler to create a base. I added the flowers at different angles and heights and set aside a few of the showiest blooms that I wanted to highlight, then threaded them in at the very end.

INGREDIENTS   Dahlia 'Alauna Pochette Surprise' • Dahlia 'Asahi Chohji' •
Dahlia 'Bon Odori' • Dahlia 'Born Rojo' • Dahlia 'Fire and Ice' • Dahlia 'Friquolet' •
Dahlia 'Myrtle's Brandy' • Dahlia 'Santa Claus' • Dahlia 'Windmill'

116

DAHLIA
'ASAHI CHOHJI'

DAHLIA
'FRIQUOLET'

DAHLIA
'MYRTLE'S BRANDY'

DAHLIA
'WINDMILL'

DAHLIA 'ALAUNA
POCHETTE SURPRISE'

DAHLIA
'BON ODORI'

DAHLIA
'FIRE AND ICE'

DAHLIA
'BORN ROJO'

DAHLIA 'SANTA CLAUS'

# AUTUMN HARVEST

For me, there's no better way to usher in autumn than filling a giant crock with fruiting branches, autumn leaves, and my favorite sunset-toned dahlias. As the nights grow cooler and the end of the flower season is on the horizon, I find that dahlias are at their peak, and I take every opportunity to soak up their magic before they are gone for the year. The cooler temps and decreasing daylight trigger the plants in the garden to start changing color, and before I know it, the whole landscape is a riot of brilliant fall tones.

Because I was using such large, heavy ingredients in this display, it was important to choose a heavy, sturdy vase that wouldn't tip over due to the ingredients' weight. I filled an old, chipped brown crock that I found at a local antique store with a half dozen fruit-laden crabapple branches that I removed the leaves from in order to expose the fruit. Then I added brilliantly colored maple leaves, and once the framework was established, I placed a number of large nodding blooms of 'Sierra Glow' dahlias around the edge of the crock, making sure to twist and turn their flower heads so they weren't all facing forward. Then I slipped in the nodding, spiky blooms of 'Bed Head' dahlias and threaded in smaller dahlia varieties to fill the spaces in between. I finished the arrangement by sprinkling miniature rose hips and golden currant foliage into all the empty cracks to add a sparkling quality to the finished arrangement.

INGREDIENTS   Amur maple foliage • Beauty bush (*Kolkwitzia amabilis*) • Crabapple 'Evereste' • Dahlia 'Bed Head' • Dahlia 'Cornel Bronze' • Dahlia 'Hy Patti' • Dahlia 'Irish Glow' • Dahlia 'Rose Toscano' • Dahlia 'Sierra Glow' • Golden currant foliage • Rose hips from *Rosa* 'Dupontii'

DAHLIA 'HY PATTI'

DAHLIA 'BED HEAD'

DAHLIA 'CORNEL BRONZE'

BEAUTY BUSH
(*KOLKWITZIA
AMABILIS*)

AMUR MAPLE
FOLIAGE

GOLDEN CURRANT
FOLIAGE

DAHLIA
'ROSE TOSCANO'

DAHLIA 'SIERRA
GLOW'

CRABAPPLE
'EVERESTE'

DAHLIA
'IRISH GLOW'

ROSE HIPS FROM
*ROSA* 'DUPONTII'

# SHOOTING STARS

I have a soft spot for open-centered dahlia varieties for two reasons. One, the bees love them, and two, they have so much personality—it's almost like their little faces are looking at you in the garden. While out in our trial patch, I found myself coming back to a small section of dainty white varieties and decided they deserved an arrangement of their own. I snipped a big handful and as I was walking back to the studio stopped off for some variegated mint and noticed the silene Blushing Lanterns and phlox 'Whipped Cream' blooming nearby. All three ingredients offered a great base in which to nestle the cheerful white dahlias.

For this easy-to-put-together posy, I chose a petite trophy-shaped ceramic vase created by my friend and famed East Coast potter Frances Palmer. Using her vases is always such a treat and makes everything look good. I started by making a simple grid using variegated mint, and placed silene between the mint stems. I then added a few little clouds of phlox, and threaded in the dahlias at varying heights, making sure to twist and turn them so that their full personality could be seen and enjoyed.

INGREDIENTS   Dahlia 'Ferncliff Alpine' • Dahlia 'Little Snowdrop' • Dahlia 'Star Child' • Dahlia 'Verrone's Morning Star' • Phlox 'Whipped Cream' • Pineapple mint • Silene Blushing Lanterns

DAHLIA 'FERNCLIFF ALPINE'

PINEAPPLE MINT

DAHLIA 'VERRONE'S MORNING STAR'

DAHLIA 'STAR CHILD'

DAHLIA 'LITTLE SNOWDROP'

SILENE BLUSHING LANTERNS

PHLOX 'WHIPPED CREAM'

# CHAMPAGNE TOAST

In recent years, the most popular wedding colors have been blush and champagne. Most of the floral designers I know are ready for this trend to pass, but I hope it's here to stay, since I love these shades. While there are thousands of dahlia varieties to choose from, this color class is the smallest and most underrepresented, probably due to the fact that their subtle and nuanced coloring doesn't fit neatly into many dahlia societies' standard grading criteria. Many of the dahlia hybridizers I've talked to think these colors are hideous, which blows my mind and just goes to show that beauty is definitely in the eye of the beholder.

This arrangement features three of my all-time favorite varieties: 'Break Out', 'Appleblossom', and 'Maya'. I chose a simple footed glass urn so as not to compete with the flowers. I used 'Break Out' to create the base of the bouquet, then added a dozen branching stems of 'Maya' and threaded in 'Appleblossom', making sure to keep its flowers above the others so that they could sway in the breeze. As a finishing touch, I took hop vines and removed all of the leaves so that the apple-green bracts were exposed and wove them along the outer edges of the arrangement. I personally hope that this palette stands the test of time and that I can even expand the range of blooms in this colorway through our farm's hybridizing program.

INGREDIENTS   Dahlia 'Appleblossom' • Dahlia 'Break Out' • Dahlia 'Maya' • Hops 'Cascade'

DAHLIA
'APPLEBLOSSOM'

DAHLIA 'MAYA'

HOPS 'CASCADE'

DAHLIA
'BREAK OUT'

# PEACH SORBET

Our dahlia field is planted in rainbow order, and when you see all of the varieties together, it's very clear which colors I'm drawn to the most. Colors in sunset tones are by far the most dominant, and even with so many wonderful varieties to choose from, I still find myself adding new varieties to the must-grow list in this palette every year.

While I love harvesting and arranging flowers from the garden, there's nothing that makes me as happy as giving flowers away. This sweet little posy was made as a gift for a friend who just moved back to town after a long time away. I found a small galvanized pail and filled it with autumn-tinted snowberry, currant, and peony foliage along with a few stems of crabapples (with the leaves removed) to create a base for the flowers amongst the leaves. I nestled in 'Cupcake' and 'Maya' dahlias, varying their stem length to give the bouquet some added dimension, then tucked in a few sprays of my favorite chrysanthemum among the other blooms. Lastly, I threaded in some miniature rose hips to echo the fall colors.

There's nothing more rewarding than sharing the abundance from the garden, especially with those you love.

INGREDIENTS   Chrysanthemum 'Pat Lehman' • Crabapple 'Evereste' • Dahlia 'Cupcake' • Dahlia 'Maya' • Golden currant foliage • Peony foliage • Rose hips from *Rosa* 'Darlow's Enigma' and 'Dupontii' • Snowberry foliage

ROSE HIPS FROM *ROSA* 'DARLOW'S ENIGMA'

GOLDEN CURRANT FOLIAGE

CHRYSANTHEMUM 'PAT LEHMAN'

DAHLIA 'CUPCAKE'

ROSE HIPS FROM *ROSA* 'DUPONTII'

DAHLIA 'MAYA'

SNOWBERRY FOLIAGE

PEONY FOLIAGE

CRABAPPLE 'EVERESTE'

# CHAPTER SIX

# VARIETY FINDER

We grew more than 800 varieties in our fields when we were creating this book, and countless more exist throughout the world. While I have many that I love, we narrowed down our top picks to 360 dahlias to include in this directory. Those listed here are all ones we grow and love, and we chose them because of their beauty and versatility when arranging.

Many of the dahlias you'll see here are commonly available to home gardeners, while others are a bit trickier to find. Our original stock of the more rare varieties was sourced from dozens of breeders and small specialty growers throughout North America and countless hours scouring the Internet over several years. The intention behind featuring some of the harder-to-find varieties wasn't to frustrate readers, but to ultimately help shine a spotlight on these treasures and increase awareness in the hopes that they make it into commercial production and aren't lost over time.

When I'm creating bouquets, I usually look for ingredients based first on color, then on form and size within that. With that in mind, I organized this dahlia directory into sections by color, loosely grouping the varieties into what I've defined as 11 color classes. You'll see a lot of variation within each color, and our categorization may differ from the way other dahlia growers and societies organize these varieties. My interpretation is based on what they look like in our fields and how they are described in the floral and design world. I then listed varieties in alphabetical order within each color and included the form, bloom size, and any special notes about each variety.

We trialed each and every dahlia in this guide, and I recommend all of them for cutting. We weeded out the weak performers and finicky varieties with wimpy stems or a trait like their heads popping off easily. Within the descriptions, we do note some that are especially good for wedding work, market sales, and other uses.

A few notes about size: In general, dahlias can range in height from 15 in (38 cm) to more than 6 ft (1.8 m) tall. We don't list specific plant heights for each variety since those can vary widely depending on your growing region, your soil, the sun exposure in your garden, and the amount of water your plants get. We use the same growing methods (detailed on page 32) for every dahlia that we grow, and we find these equally effective for all of them. For the varieties listed ahead, we do note some that are particularly suited to growing in a large container or a mixed garden border because they have a more contained growth habit.

Similarly, bloom sizes are highly variable. For example, 'Café au Lait' flowers can reach 10 in (25 cm) across on our farm, but in a friend's California garden they're consistently 6 in (15 cm). The measurements we list for each variety come from our own research as well as that of other growers and breeders, and we include them to give you a general idea of how big your flowers may become. But keep in mind that the blooms of any given variety may be larger or smaller than what's listed, based on all the conditions of the garden in which it grows.

Lastly, "Floret's Favorites," marked with a ♥, are varieties that we just can't live without. To find suppliers, see Resources, page 219.

# WHITE

Timeless and classically appealing, this category is
made up of bright clean shades of white, warm ivory,
cool silver, pure snow, and fresh cream. Not surprisingly, these varieties are popular for weddings.

## 'AC Casper'

**FORM:** Informal Decorative

**BLOOM SIZE:** 8 to 10 in (20 to 25 cm)

Large, creamy white flowers are a brilliant contrast to glowing green foliage. These sturdy growers are a fantastic size and shape for wedding work.

## 'Allie White'

**FORM:** Informal Decorative

**BLOOM SIZE:** 6 to 8 in (15 to 20 cm)

Flowers are held on strong stems, and they are a wonderful shape and size. Its soft appearance and lovely color make it ideal for arranging.

## 'Andrea Lawson'

**FORM:** Ball

**BLOOM SIZE:** 2 to 4 in (5 to 10 cm)

This tall, vigorous grower has crisp white blooms that have a hint of lavender pink in the center and on the tips. These are ideal for hand-tied bouquets.

## 'Blizzard' ♥

**FORM:** Formal Decorative

**BLOOM SIZE:** Up to 4 in (10 cm)

A vigorous grower, this variety produces masses of lush green foliage. Each plant is loaded with an abundance of blooms borne on strong stems. The flowers are a great all-around size.

## 'Boom Boom White'

**FORM:** Formal Decorative

**BLOOM SIZE:** 3 to 4 in (8 to 10 cm)

Large plants have creamy white ball-shaped flowers on tall strong stems. Great for wedding work.

## 'Bowen'

**FORM:** Pompon

**BLOOM SIZE:** Up to 2 in (5 cm)

Adorable button blooms are white with the occasional blush. They're perfect for bridal bouquets.

'Bride to Be'

**FORM:** Waterlily

**BLOOM SIZE:** 4 in (10 cm)

These clean, white blooms are abundant on medium-size plants. Because of their upward-facing flowers, they are one of the best choices for arranging.

'Center Court'

**FORM:** Formal Decorative

**BLOOM SIZE:** 6 to 8 in (15 to 20 cm)

Producing pure snow-white flowers atop long, strong stems, this variety is especially vigorous. Slightly upward-facing blooms are lovely for straight bunches and bouquets.

'Colwood Hope'

**FORM:** Laciniated

**BLOOM SIZE:** 6 to 8 in (15 to 20 cm)

Laciniated creamy white blooms make this a winner in our dahlia patch. Rich green foliage with deeply serrated edges contributes to its visual impact.

'Corn Bride'

**FORM:** Formal Decorative

**BLOOM SIZE:** 6 to 8 in (15 to 20 cm)

These star-like soft white flowers have a shimmery effect. Long stems are perfect for arranging.

'Creamy' ♥

**FORM:** Mini Ball

**BLOOM SIZE:** 2 to 3½ in (5 to 9 cm)

Rounded blooms are the color of fresh cream. This incredible variety is wonderful for bouquets and makes a lovely garden addition.

'Dorothy R'

**FORM:** Mini Ball

**BLOOM SIZE:** 2 to 3½ in (5 to 9 cm)

Ivory blooms with lime-green centers grow on medium plants that have vibrant green foliage. This clean crisp white is perfect for wedding work.

### 'Ferncliff Alpine' ♥

**FORM**: Collarette

**BLOOM SIZE**: 4 in (10 cm)

Tall, vigorous plants are loaded with the most beautiful white star-shaped blooms that have glowing golden centers on long, strong stems. This is one of our favorites of the white single-flowered forms.

### 'Ferncliff Pearl' ♥

**FORM**: Formal Decorative

**BLOOM SIZE**: 4 in (10 cm)

White blooms with faint iridescent blush centers look just like pearls. This medium, vigorous grower has deep green foliage and is one of the best white dahlias.

### 'Hollyhill Miss White'

**FORM**: Mini Ball

**BLOOM SIZE**: 2 to 3½ in (5 to 9 cm)

Towering plants produce a sea of perfectly formed clear white flowers, each with the palest lavender center. This is one of the best white mini balls you can grow.

### 'Kenora Challenger'

**FORM**: Semi-Cactus

**BLOOM SIZE**: 6 to 8 in (15 to 20 cm)

Medium-size plants have perfectly formed starburst-like blooms. This well-loved variety is a standard seed parent for breeding.

### 'L'Ancresse'

**FORM**: Ball

**BLOOM SIZE**: Over 3½ in (9 cm)

One of the very best white ball varieties available, this treasure churns out perfect flowers all season long. Its long, strong stems make it an excellent cut flower. Great for wedding work and hand-tied bouquets.

### 'Lulu Island Art'

**FORM**: Formal Decorative

**BLOOM SIZE**: 6 to 8 in (15 to 20 cm)

Snow-white flowers top long, strong stems that are great for cutting. Perfect for event work, its shape makes it stunning in the garden.

### 'Narrows Ryder'

**FORM:** Formal Decorative

**BLOOM SIZE:** 4 to 6 in (10 to 15 cm)

Creamy white flowers that look like fluffy clouds have the palest lavender centers. Plants are medium size, and blooms are great for weddings.

### 'Platinum Blonde' ♥

**FORM:** Anemone

**BLOOM SIZE:** 4 in (10 cm)

One of the most unusual varieties we grow; the flowers resemble double-flowered echinacea. Each fuzzy buttercream center on this bloomer is surrounded by a ring of bright white petals. Loved by designers, the long-stemmed flowers are great for arranging and last well in the vase.

### 'RJR' ♥

**FORM:** Formal Decorative

**BLOOM SIZE:** 4 to 6 in (10 to 15 cm)

This is one of the best white varieties you can grow. Tall plants produce an abundance of long-stemmed pure white flowers that are a great size for arranging.

146

### 'R Kris'

**FORM:** Informal Decorative

**BLOOM SIZE:** 6 to 8 in (15 to 20 cm)

Reflexed petals are creamy white with a greenish cast. Flowers have a feathery, windswept appearance.

### 'Robann Pristine'

**FORM:** Formal Decorative

**BLOOM SIZE:** 4 to 6 in (10 to 15 cm)

A vigorous grower, these tall plants are topped with medium-size, round blooms. Petals show the palest touch of purple and are a romantic addition to wedding work.

### 'Ryecroft Brenda T'

**FORM:** Formal Decorative

**BLOOM SIZE:** 4 to 6 in (10 to 15 cm)

Dome-shaped blooms have light lavender centers that lend themselves well to bouquets. Plants are medium size and productive.

'Small World' ♥

**FORM:** Pompon

**BLOOM SIZE:** Up to 2 in (5 cm)

One of the best white dahlias we've ever grown. It's smothered in flowers all season long. Strong stems and weather-proof blooms make it excellent for cutting. It's especially prized by wedding florists.

'Snowbound'

**FORM:** Informal Decorative

**BLOOM SIZE:** 8 to 10 in (20 to 25 cm)

These blooms have a soft, downy quality, and petals that curve at their tips further accent that beauty. Medium-size plants churn out flowers.

'Sterling Silver'

**FORM:** Formal Decorative

**BLOOM SIZE:** 6 to 8 in (15 to 20 cm)

This beautiful, full, warm white variety has slightly pointed petals that reflex, making it a real standout in arrange-ments. Plants are vigorous and tall, and flowers are held on long stems.

'White Aster'

**FORM:** Pompon

**BLOOM SIZE:** 2 to 2½ in (5 to 6 cm)

Tall strong stems carry small ivory rounded blooms. This productive variety is great for bouquets.

'White Nettie'

**FORM:** Mini Ball

**BLOOM SIZE:** 2½ in (6 cm)

This darling is a fantastic size for arrang-ing and produces lots of flowers. It's great for wedding work.

'Wyn's Ghostie'

**FORM:** Informal Decorative

**BLOOM SIZE:** 6 to 8 in (15 to 20 cm)

This strong grower reminds us of swan feathers, with petals that are reflexed and twist at the tips. Long, strong stems support these incredibly unique blooms.

# YELLOW

This group includes one of the most diverse ranges of colors we show. It includes buttercream, lemon meringue, canary, goldenrod, honey, and highlighter yellow, which has a fluorescent quality and slight green undertone. This color range is by far the most cheerful.

### 'Alpen Sundown'

**FORM:** Formal Decorative

**BLOOM SIZE:** 4 to 6 in (10 to 15 cm)

This beautiful treasure has exquisite coloring that stands out in the garden. Soft butter-yellow petals are delicately outlined and streaked with raspberry. While plants are on the petite side, this variety produces long stems that are excellent for cutting.

### 'Big Brother'

**FORM:** Informal Decorative

**BLOOM SIZE:** 8 to 10 in (20 to 25 cm)

We have been on the quest to find the perfect large-flowered golden dahlia for years and happily discovered this magnificent gem during a recent variety trial. The ruffled tawny blooms are wonderful in large-scale arrangements.

### 'Blah Blah Blah' ♥

**FORM:** Formal Decorative

**BLOOM SIZE:** 4 to 6 in (10 to 15 cm)

Floral designers love the standout color of this variety. Blooms are butterscotch with peach-lavender centers. This beauty combines easily with many hues.

### 'Bloomquist Butter Cream' ♥

**FORM:** Informal Decorative

**BLOOM SIZE:** 4 to 6 in (10 to 15 cm)

Soft buttercream flowers with a slight hint of blush have glowing centers. Stems are long and strong. Beautiful combined with both yellow and blush tones.

### 'Blyton Softer Gleam'

**FORM:** Ball

**BLOOM SIZE:** 5 in (13 cm)

Flowers are the loveliest shade of soft gold and resemble wheat straw brushed with a hint of soft orange on the petal tips. An extremely prolific bloomer; the flowers are long lasting and weather resistant.

### 'Born Sty' ♥

**FORM:** Stellar

**BLOOM SIZE:** 6 to 8 in (15 to 20 cm)

When we think of the best yellow dahlias we've grown, this is certainly on the list. Long, graceful stems display flowers the shade of pale lemon ice with the softest peach centers. It's absolutely irresistible.

### 'Bracken Sarah' ♥

**FORM:** Formal Decorative

**BLOOM SIZE:** 6 to 8 in (15 to 20 cm)

The golden-peach blooms of this beauty are held upright on sturdy stems. Plants are towering, and the flowers are great in hand-tied bouquets.

### 'Buttercup' ♥

**FORM:** Pompon

**BLOOM SIZE:** Up to 2 in (5 cm)

Clear, cheerful yellow blooms look like glowing orbs on the tips of the stems. Foliage is a contrasting dark green, and vigorous plants are very productive.

### 'CG Amber'

**FORM:** Formal Decorative

**BLOOM SIZE:** 4 to 6 in (10 to 15 cm)

These petite plants with glossy green foliage have such a fitting name. Blooms are peachy-orange with darker amber centers.

### 'Cherish'

**FORM:** Formal Decorative

**BLOOM SIZE:** Up to 4 in (10 cm)

These compact plants are good bloomers and an excellent addition to the garden. Petals are a soft buttercream hue with lilac veining and centers. The unique coloring, paired with strong stems, make this a great choice for arranging.

### 'Chimacum Wendy'

**FORM:** Ball

**BLOOM SIZE:** 4 to 6 in (10 to 15 cm)

Large, soft goldenrod yellow flowers have a slightly darker orange dusting on the petal tips. This is a reliable, productive variety.

### 'Citron du Cap'

**FORM:** Laciniated

**BLOOM SIZE:** 6 to 8 in (15 to 20 cm)

Petals that are split at the tips give this variety a feathery quality. Pale buttercream petal ends are touched with blush, and the overall color lightens as flowers age. One of the most romantic yellow varieties we grow.

### 'Clearview Citron'

**FORM:** Incurved Cactus

**BLOOM SIZE:** 4 to 6 in (10 to 15 cm)

Flowers are the color of lemon ice, and pointed petals have the softest blush dusting in the center. Tall, vigorous plants produce long, strong stems.

### 'Daisy Mae'

**FORM:** Anemone

**BLOOM SIZE:** 4 to 6 in (10 to 15 cm)

Sunshine-colored blooms have distinct shaggy centers. Medium-size plants have lime-green streaked stalks that carry the nodding flower heads.

### 'Formby Crest'

**FORM:** Formal Decorative

**BLOOM SIZE:** 3 to 4 in (8 to 10 cm)

The rounded flower heads of this beauty are a glowing gold. Blooms are weather resistant, and plants produce long stems all season long.

### 'Golden Scepter'

**FORM:** Formal Decorative

**BLOOM SIZE:** 2½ in (6 cm)

The large, vigorous plants of this cheerful variety produce an abundance of small, tangerine flowers all season on long stems. The petite blooms are perfect for mixed bouquets and wedding work.

### 'Hamari Gold'

**FORM:** Informal Decorative

**BLOOM SIZE:** 8 to 10 in (20 to 25 cm)

This giant-flowered gem is a beautiful warm bronzy-gold and is always a favorite with everyone who visits our garden. Its blooms mix beautifully with just about anything, and plants are especially striking in the fall.

### 'Happy Butterfly'

**FORM:** Informal Decorative

**BLOOM SIZE:** 4 in (10 cm)

Tall plants are loaded with the prettiest butterfly-shaped blooms in a mix of soft yellow and cranberry, with dark reverse petals. The upward-facing flowers have long, strong stems, making them great for cutting and arranging.

### 'Hollyhill Calico'

**FORM:** Mini Ball

**BLOOM SIZE:** 2 to 3½ in (5 to 9 cm)

This darling treasure was a standout in our trial patch. The tricolor blooms are a distinctive mix of peach, white, and lemon, and no two flowers are alike. An excellent cut flower and garden addition.

### 'Hollyhill Frodo'

**FORM:** Pompon

**BLOOM SIZE:** Up to 2 in (5 cm)

Blooms are goldenrod with a light dusting of cranberry on the tips and back of petals giving them a metallic quality. Simply adorable, these flowers are well suited for boutonnieres or bouquets.

### 'Honeydew'

**FORM:** Formal Decorative

**BLOOM SIZE:** 5 to 7 in (13 to 17 cm)

Delightful soft peach blooms have dominant yellow centers and highlights that glow in the garden. This versatile color is great for arranging.

### 'Irish Pinwheel' ♥

**FORM:** Stellar

**BLOOM SIZE:** 4 to 6 in (10 to 15 cm)

One of the coolest varieties in the dahlia patch! Beautiful, uniquely pointed petals are a soft gold, brushed with a hint of cherry red. Plants churn out an abundance of tall, strong stems all season long, making them a fantastic variety for cutting.

### 'Karmel Korn' ♥

**FORM:** Informal Decorative

**BLOOM SIZE:** 6 in (15 cm)

This is one of the best yellow varieties we grow. Warm buttercream blooms with purple undertones have the lightest lavender blush in the center and are versatile in bouquets. They remind us of tissue paper flowers.

### 'Lakeview Peach Fuzz' ♥

**FORM:** Laciniated

**BLOOM SIZE:** 4 to 6 in (10 to 15 cm)

With an extremely versatile color range, this variety combines well with lots of tones. Some blooms are a soft honey, while others lean to apricot and yellow. Petals have laciniated tips, giving a shaggy appearance. Long, willowy stems hold these all-time favorite flowers.

### 'Lucky Ducky' ♥

**FORM:** Anemone

**BLOOM SIZE:** 3 in (8 cm)

This is one of my favorites in this color family. The center of the flower is dense and fuzzy, and it is encircled by clean, clear yellow petals whose outer edges are tipped in white. Extremely long lasting for a single-flowered variety. Good in pots.

### 'Mary Lou'

**FORM:** Formal Decorative

**BLOOM SIZE:** 5 to 6 in (13 to 15 cm)

This tall golden beauty has upward-facing blooms with long stems that make excellent cut flowers. Egg yolk–yellow blooms glow in the garden.

### 'Miss Amara'

**FORM:** Formal Decorative

**BLOOM SIZE:** 4 to 6 in (10 to 15 cm)

This peachy-yellow beauty is a favorite at our farm. The vigorous plants produce an abundance of cheerful, long-stemmed blooms that are perfect for flower arranging.

### 'MM Buttercream'

**FORM:** Ball

**BLOOM SIZE:** Over 3½ in (9 cm)

Of all the ball dahlias we've ever seen, this is one of the most beautiful. Perfectly shaped pale buttercream blooms are held on vigorous, long, strong stems.

### 'Ms Kennedy'

**FORM:** Mini Ball

**BLOOM SIZE:** 2 to 3½ in (5 to 9 cm)

A true garden workhorse, this variety churns out small golden flowers with adorable orange button centers that are a great size for cutting. Long, strong stems and good weather resistance make this variety a must-grow.

### 'Ms Lisa' ♥

**FORM:** Informal Decorative

**BLOOM SIZE:** 6 to 8 in (15 to 20 cm)

Dense flower heads are warm peachy gold with lighter tips, and back petals have a glittery iridescent quality. The pointed petals twist and twirl like feathers.

### 'Oreti Adele'

**FORM:** Formal Decorative

**BLOOM SIZE:** 4 to 6 in (10 to 15 cm)

One of the most beautiful varieties we grow, this treasure is a lovely mix of peach and warm gold. Flowers change color slightly depending on the weather.

### 'Redhawk Buckskin' ♥

**FORM:** Formal Decorative

**BLOOM SIZE:** 5½ in (14 cm)

These medium-size plants are smothered in soft gold blooms that have the tiniest hint of blush. This rare variety is hard to find but is one of my all-time favorites.

### 'Sandia Brocade' ♥

**FORM:** Anemone

**BLOOM SIZE:** 3 to 4 in (8 to 10 cm)

Contrasting against dark stems, blooms are the shade of cantaloupe. Outer petals reflex back, while the center is a textural golden pouf with layers of tubular fringed petals. Holds up well in poor weather despite its delicate appearance.

### 'Sandia Sunbonnet'

**FORM:** Anemone

**BLOOM SIZE:** 4 in (10 cm)

Cheerful clear lemon blooms have egg yolk–yellow centers and long, thin stems. Its glowing color radiates from the garden and the vase.

### 'Skipley Moonglow' ♥

**FORM:** Ball

**BLOOM SIZE:** Over 3½ in (9 cm)

Plants are on the shorter side and produce the prettiest pale buttercream flowers that have a light blush, giving them a glowing quality. Their tones make them great for wedding work.

### 'Sugar Daddy'

**FORM:** Formal Decorative

**BLOOM SIZE:** 2 in (5 cm)

The sweetest petite, rounded blooms remind us of butterscotch candy. This garden workhorse blooms on long, strong stems. A great variety for bouquets.

### 'Suncrest' ♥

**FORM:** Ball

**BLOOM SIZE:** 4 in (10 cm)

After trialing hundreds of yellow varieties over the years, this beauty is still a top
favorite. Soft canary-yellow blooms have
a faint apricot on the petal tips. Productive, long-lasting, and weather-resistant
flowers make it perfect for market sales.

### 'Valley Tawny' ♥

**FORM:** Formal Decorative

**BLOOM SIZE:** 4 to 6 in (10 to 15 cm)

Breeders David and Leone Smith were
generous enough to share this treasure
with us. The butterscotch-colored blooms
of this stunning variety are borne in great
abundance all season long and make great
cut flowers. Plants are hardy, weather
resistant, and free flowering.

### 'Verrone's Richard B'

**FORM:** Stellar

**BLOOM SIZE:** 4 to 6 in (10 to 15 cm)

Petite, whorled blooms are honey colored
with pale orange backing on tall stems.
Out in the patch this variety looks like
glowing pinwheels.

### 'Westerton Lillian' ♥

**FORM:** Informal Decorative

**BLOOM SIZE:** 6 to 8 in (15 to 20 cm)

Massive, towering plants are loaded
with pale lemon blooms that have blush
centers and strong stems. Its soft appearance makes it well suited for wedding
work.

### 'Winholme Diane'

**FORM:** Formal Decorative

**BLOOM SIZE:** 4 to 6 in (10 to 15 cm)

Towering plants have strong stems topped
with beautiful flowers that remind us
of lemon shaved ice. Bloom centers are
gently kissed with blush, making it a
standout for arranging.

### 'Wyn's New Pastel'

**FORM:** Formal Decorative

**BLOOM SIZE:** 6 in (15 cm)

A pretty new addition to our farm, this
stunner is golden yellow mixed with
peachy blush. Long, strong stems are
borne in abundance on this hardy, vigorous grower.

# BLUSH/CHAMPAGNE

Subtle and romantic, this group includes both soft pink and warm neutral tones that have an effervescent quality, including blush, rose, and champagne all the way to buff, beige, and nude. This color category is the most sought after for weddings.

**'Appleblossom'** ♥

**FORM:** Collarette

**BLOOM SIZE:** 4 in (10 cm)

One of the most popular varieties we grow, it's a hit with event designers and makes a wonderful wedding bouquet addition. The petals start out a soft buttercream and transform to a delicate blush.

**'Apricot Star'**

**FORM:** Cactus

**BLOOM SIZE:** 6 in (15 cm)

This upright grower has star-shaped flowers in a mix of soft gold and champagne. Branching stems produce loads of textural blooms.

**'April Heather'** ♥

**FORM:** Collarette

**BLOOM SIZE:** 4 in (10 cm)

Beautiful gold and buff flowers are great for wedding work. This is one of the most neutral-toned dahlias, and it can go pink or beige. Petals are very long lasting and fade to a beautiful buff as they age.

**'Break Out'** ♥

**FORM:** Informal Decorative

**BLOOM SIZE:** 8 to 10 in (20 to 25 cm)

One of the loveliest varieties we have ever grown, this large-flowered treasure produces an abundance of soft warm pink blooms brushed with buttercream on long, strong stems.

**'Café au Lait'** ♥

**FORM:** Informal Decorative

**BLOOM SIZE:** 8 to 10 in (20 to 25 cm)

This is our most requested variety. The massive blooms resemble pastel silk pillows. Popular with brides and wedding designers, the flowers are the most unusual shade of pale creamy blush.

**'Camano Zoe'** ♥

**FORM:** Mini Ball

**BLOOM SIZE:** 2 to 3½ in (5 to 9 cm)

This adorable variety boasts strong, upright growth and super-soft, blush-pink blooms. A weather-resistant variety, it makes an excellent choice for wedding work.

### 'Castle Drive' ♥

**FORM:** Formal Decorative

**BLOOM SIZE:** 4 to 6 in (10 to 15 cm)

One of the first varieties to flower each summer in the garden, this variety produces a bumper crop of the most beautiful soft blush-pink, medium-size blooms that are great for arranging and wedding work.

### 'Chimacum Davi'

**FORM:** Mini Ball

**BLOOM SIZE:** 2 to 3½ in (5 to 9 cm)

The Smiths, some of our favorite dahlia breeders, lovingly created this special variety and were generous enough to share planting stock with us. Tall, strong stems carry pink blooms with darker centers.

### 'Clearview Palser'

**FORM:** Informal Decorative

**BLOOM SIZE:** 8 to 10 in (20 to 25 cm)

Blooms have lighter golden centers and tips edged in lavender that give an overall buff appearance. This versatile coloring combines well with many shades.

### 'Crazy 4 Teedy'

**FORM:** Formal Decorative

**BLOOM SIZE:** 4 to 6 in (10 to 15 cm)

These blooms are a unique blend of yellow and pink. Weatherproof flowers are held on long, strong stems and make long-lasting cut flowers.

### 'Cupcake' ♥

**FORM:** Formal Decorative

**BLOOM SIZE:** 4 to 6 in (10 to 15 cm)

This introduction from Swan Island is a stocky grower with blooms that have one-of-a-kind blended color. These long-lasting cut flowers are a must-grow for wedding work and arranging.

### 'Day Dreamer'

**FORM:** Waterlily

**BLOOM SIZE:** 4 in (10 cm)

These look just like waterlilies on a pond. Cheerful blooms are glowing apricot-peach with pale lemon centers.

### 'Diana's Memory'

**FORM**: Waterlily

**BLOOM SIZE**: Up to 4 in (10 cm)

Beautiful creamy champagne flowers have a green undertone and lavender tips on long stems. Plants are on the shorter side and great for growing in pots.

### 'Fairway Pilot'

**FORM**: Formal Decorative

**BLOOM SIZE**: 10+ in (25+ cm)

This giant of the garden is a stocky grower, producing immense flowers with huge, soft salmon-blush petals. A spectacular choice for event work. The blooms are somewhat fragile but worth growing for their beauty.

### 'Ferncliff Dolly' ♥

**FORM**: Collarette

**BLOOM SIZE**: 4 to 6 in (10 to 15 cm)

The warm pink petals have softer blush tips and a ring of smaller creamy inner petals around a glowing golden center. Long-lasting blooms fade beautifully.

### 'Hapet Pearl'

**FORM**: Mini Ball

**BLOOM SIZE**: 2 to 3½ in (5 to 9 cm)

Buttercream flowers are dusted in raspberry and have darker centers. Medium-size plants produce glossy green foliage and tall stems that are great for arranging.

### 'Honka Fragile' ♥

**FORM**: Orchid

**BLOOM SIZE**: 2 to 3 in (5 to 8 cm)

Tall, vigorous plants are loaded with the most eye-catching cranberry-edged, white star-shaped flowers. A long-lasting cut flower, these beauties almost don't look real.

### 'Innocence'

**FORM**: Waterlily

**BLOOM SIZE**: 6 in (15 cm)

Pale blush-pink blooms have creamy white centers. Plants produce an abundance of blooms on strong stems. Perfect for wedding work.

### 'KA's Cloud' ♥

**FORM:** Informal Decorative

**BLOOM SIZE:** 8 to 10 in (20 to 25 cm)

This stunning introduction from Santa Cruz Dahlias has fast become a favorite. The massive, blush-kissed white blooms produce abundantly on long stems all season long. A must-have for wedding work!

### 'Kelgai Ann'

**FORM:** Waterlily

**BLOOM SIZE:** 5 in (13 cm)

This treasured variety produces large upward-facing saucer-shaped flowers that are the softest blush pink with darker pink streaks. It's delicate and also striking.

### 'Maltby Pearl'

**FORM:** Formal Decorative

**BLOOM SIZE:** 4 in (10 cm)

The soft, creamy pink petals of this beauty are edged with lavender and tipped with white. This unusual color combination gives them an almost metallic effect. They're fantastic for bouquets and arranging.

### 'Maya' ♥

**FORM:** Formal Decorative

**BLOOM SIZE:** 5 to 6 in (13 to 15 cm)

Strong stems have upward-facing flowers in glowing champagne with blush tips. Flowers are often borne in sprays, which is an unusual quality for a dahlia but makes them excellent for flower arranging.

### 'Miracle Princess'

**FORM:** Waterlily

**BLOOM SIZE:** 4 to 6 in (10 to 15 cm)

Soft-pink-and-white flowers have a really sweet quality. Vigorous plants are productive, and long stems have a slightly dark tint.

### 'Nadia Ruth'

**FORM:** Laciniated

**BLOOM SIZE:** 6 to 8 in (15 to 20 cm)

Blooms are a mix of shell pink and buttercream with petal tips that have a feather-like quality. This variety flowers abundantly.

'Normandy Deegee'

**FORM:** Formal Decorative

**BLOOM SIZE:** 4 to 6 in (10 to 15 cm)

This medium-size plant has glowing buttercream blooms edged in lavender pink, which gives the flowers lots of dimension. Great for wedding work.

'Odyssey' ♥

**FORM:** Mini Ball

**BLOOM SIZE:** 2 to 3½ in (5 to 9 cm)

Adorable small blush-pink-and-cream blooms are dusted with a hint of purple. They are perfect for romantic hand-tied bouquets.

'Roque Starburst'

**FORM:** Semi-Cactus

**BLOOM SIZE:** Over 10 in (25 cm)

Massive blooms have a latte-colored center. Plants are vigorous and productive, and the flowers have a striking, eye-catching shape.

'Sheer Heaven' ♥

**FORM:** Formal Decorative

**BLOOM SIZE:** 5 in (13 cm)

This cheerful variety is a special mix of soft peach and the palest lemon yellow. Flowers are upward facing and borne on long, strong stems. Wonderful for flower arranging.

'Sidehill Trishie' ♥

**FORM:** Formal Decorative

**BLOOM SIZE:** 4 to 6 in (10 to 15 cm)

This lovely variety boasts medium-size warm buttercream yellow blooms with melon-pink undertones. Upward-facing flowers are great for arranging.

'Silver Years'

**FORM:** Waterlily

**BLOOM SIZE:** 4 in (10 cm)

The palest blush pink blooms have yellow-green centers that face up and out. This sweet and romantic variety is great for bouquets.

### 'Strawberry Ice'

**FORM:** Informal Decorative

**BLOOM SIZE:** 8 to 10 in (20 to 25 cm)

This treasure has soft pink blooms with pale yellow undertones that ride atop long, strong stems. Wonderful for wedding work and large-scale flower arranging.

### 'Susan Gillott'

**FORM:** Semi-Cactus

**BLOOM SIZE:** 4 to 6 in (10 to 15 cm)

Medium-size plants with long, strong stems have shell pink flowers and glowing centers. Great for wedding work.

### 'Sweet Nathalie' ♥

**FORM:** Formal Decorative

**BLOOM SIZE:** 5 in (13 cm)

This is one of the most beautiful varieties we've ever grown. Medium-size plants are smothered in pale blush blooms all season. Long, strong stems make this a great cut flower, perfect for romantic arrangements.

### 'Tahoma Early Dawn'

**FORM:** Informal Decorative

**BLOOM SIZE:** 4 to 6 in (10 to 15 cm)

These creamy white ruffled blooms have a feathery quality, and the petals have the faintest brush of lavender. This long-stemmed variety holds up well to weather and insects.

### 'Valley Porcupine' ♥

**FORM:** Novelty Fully Double

**BLOOM SIZE:** Up to 4 in (10 cm)

This adorable bloomer is a favorite with every floral designer who visits our farm. The petite, spiky, creamy-blush flowers are prickly to the touch, which is how it got its name. It's a fantastic, long-lasting cut flower that's perfect for arranging.

### 'Wizard of Oz'

**FORM:** Mini Ball

**BLOOM SIZE:** 2 to 3½ in (5 to 9 cm)

Medium plants are covered in soft cotton-candy pink flowers. This particular variety tends to drop its petals once flowers reach full maturity, so harvest when blooms are still young, about one-third of the way open.

# PEACH

In this category you'll find a warm blend of sherbet tones including apricot, peach, pale salmon, glowing melon, and cantaloupe. These colors are popular because of their versatility and inviting appeal.

### 'A La Mode'

**FORM:** Formal Decorative

**BLOOM SIZE:** 6 to 8 in (15 to 20 cm)

This beautiful salmon-orange and white bicolor is reminiscent of candy corn. A prolific bloomer with long stems, this customer favorite has large blooms in a range of shades and variegations.

### 'Blown Dry'

**FORM:** Informal Decorative

**BLOOM SIZE:** 6 to 8 in (15 to 20 cm)

Pale peachy-apricot blooms with golden centers give this beauty an autumnal feel, and petals look pinched and twisted at the tips. Strong stems carry these medium-size gems.

### 'Camano Sitka'

**FORM:** Incurved Cactus

**BLOOM SIZE:** 6 to 8 in (15 to 20 cm)

Large cantaloupe-colored blooms have tubular petals on large flowers that ride atop towering plants.

### 'Cara Elizabeth' ♥

**FORM:** Informal Decorative

**BLOOM SIZE:** 4 to 6 in (10 to 15 cm)

These blooms are a lovely smoky-peach hue, a color not often found. This showstopping favorite is ideal for flower arranging and wedding work and looks fantastic mixed with many colors.

### 'Clearview Peachy' ♥

**FORM:** Ball

**BLOOM SIZE:** Over 3½ in (9 cm)

Large, vigorous plants with bright green foliage are covered with pale salmon blooms that are dusted with lavender. Held on long, strong stems, each captivating flower is slightly different.

### 'Country Charm' ♥

**FORM:** Formal Decorative

**BLOOM SIZE:** 6 to 8 in (15 to 20 cm)

Glowing apricot flowers have the prettiest darker lavender center. Long, strong stems make this a particularly wonderful cut flower.

'Craig Charles' ♥

**FORM**: Informal Decorative

**BLOOM SIZE**: 4 to 6 in (10 to 15 cm)

Salmon blooms have petals that twist and wave, giving them a ruffled quality. This is the only flower with this color and shape that isn't a giant. It's the perfect size for arranging.

'Crichton Honey' ♥

**FORM**: Ball

**BLOOM SIZE**: 4 in (10 cm)

This warm apricot beauty is a favorite with designers and market customers. Blooms are weather resistant. Though plants are on the shorter side, if pinched early they will produce a bumper crop of flowers on long, strong stems that are excellent for cutting.

'Ferncliff Copper' ♥

**FORM**: Formal Decorative

**BLOOM SIZE**: 4 to 6 in (10 to 15 cm)

An absolute must-grow for flower arrangers, this versatile coppery beauty features blooms that combine well with many other hues. Plants produce a steady supply of sturdy, long-stemmed flowers all season long.

'Fiona'

**FORM**: Ball

**BLOOM SIZE**: Over 3½ in (9 cm)

Melon-colored blooms have radiating golden centers that glow in the garden and in the vase. The tight ball shape is weather resistant and great for straight bunches.

'French Doll' ♥

**FORM**: Formal Decorative

**BLOOM SIZE**: 3 in (8 cm)

A favorite with everyone who visits the garden, this beauty lives up to her name. Beautiful peach flowers have a soft yellow center and look as if they are glowing. Tall, strong stems make them excellent for cutting.

'Hamilton Lillian'

**FORM**: Formal Decorative

**BLOOM SIZE**: 4 to 5 in (10 to 13 cm)

One of the prettiest we grow in this color family, its blooms have perfectly formed pointed petals. Plants are on the shorter side, but if pinched early they will produce a bumper crop of long, strong stems that are excellent for cutting.

### 'Hapet Champagne'

**FORM:** Laciniated

**BLOOM SIZE:** 6 to 8 in (15 to 20 cm)

Beautiful glowing apricot petals surround lemony yellow centers on strong stems. With petal tips that are snipped and feather-like, there's nothing else like it.

### 'Henriette'

**FORM:** Semi-Cactus

**BLOOM SIZE:** 4 to 6 in (10 to 15 cm)

These striking flowers are the softest salmon, and dark stems offset the pastel blooms beautifully. A weather-resistant variety, this has good stems for cutting, making it an excellent choice for arranging and wedding work.

### 'Hollyhill Orange Ice'

**FORM:** Formal Decorative

**BLOOM SIZE:** 4 to 6 in (10 to 15 cm)

The unique pattern on this bicolor reminds us of candy corn. Base petals are the color of dried apricots with pale blush tips, and colors change evenly, blending together. Blooms are glittery and iridescent.

### 'Island Sunset' ♥

**FORM:** Informal Decorative

**BLOOM SIZE:** 6 to 8 in (15 to 20 cm)

This variety has every color you'd want all wrapped into one dahlia. Super-soft apricot blooms have a hint of blush, and as flowers age, they take on a pinkish cast. Curling petals have a glittering quality. A must-grow for flower arranging.

### 'Jowey Nicky' ♥

**FORM:** Ball

**BLOOM SIZE:** 2 to 4 in (5 to 10 cm)

One of the most popular varieties with floral designers, this soft melon treasure churns out an abundance of flowers all season long. Fantastic for the garden and the vase.

### 'Lark's Ebbe'

**FORM:** Formal Decorative

**BLOOM SIZE:** 4 to 6 in (10 to 15 cm)

This beautiful variety has quickly become a favorite for bouquets. The compact plants do well in borders and produce a bumper crop of peachy flowers with long stems that are perfect for arranging.

### 'L.A.T.E'

**FORM:** Formal Decorative

**BLOOM SIZE:** 4 in (10 cm)

Light peach flowers with raspberry dusting have darker pink centers. They are great for wedding work.

### 'Linda's Baby' ♥

**FORM:** Ball

**BLOOM SIZE:** 3 to 4 in (8 to 10 cm)

We have been on a quest to find the perfect peach dahlia for years, and this is everything we hoped for and more. The stunning blooms sit atop tall, strong stems, and plants are both hardy and free flowering.

### 'Linda's Esther'

**FORM:** Formal Decorative

**BLOOM SIZE:** 4 to 6 in (10 to 15 cm)

This sweet variety reminds us of vanilla-orange swirl ice cream. Some flowers are warm orange, others peachy with white tips; all have a stunning metallic quality. No two flowers are the same.

### 'Nathalie G.' ♥

**FORM:** Formal Decorative

**BLOOM SIZE:** 4 in (10 cm)

The melon-pink rounded blooms of this lovely variety have a light lavender dusting. Plants are on the shorter side, but what it lacks in size it makes up for in color and productivity.

### 'Papageno'

**FORM:** Informal Decorative

**BLOOM SIZE:** 10 to 11 in (25 to 28 cm)

Medium-size plants are smothered in large, distinctive smoky-peach blooms with lemon-yellow accents. This dinner plate is a must-grow for floral design and wedding work.

### 'Peaches N' Cream' ♥

**FORM:** Formal Decorative

**BLOOM SIZE:** 5 in (13 cm)

A truly exquisite color pattern, this bloomer has soft salmon petals that fade to white at the tips. The effect is absolutely stunning and makes it one of the most popular varieties in our garden. This variety thrives in heat and humidity.

'Rock Run Ashley' ♥

FORM: Formal Decorative

BLOOM SIZE: Up to 4 in (10 cm)

One of the most beautiful and versatile varieties from our trials, this produces a steady stream of small buff to blush blooms all season long. Plants are on the shorter side, but if pinched early they will produce long stems abundantly. Good in pots.

'Shaggy Chic'

FORM: Laciniated

BLOOM SIZE: 6 to 8 in (15 to 20 cm)

Towering plants are topped with large pinkish-coral blooms with yellow undertones giving flowers a smoky appearance. Petals are quilled and laciniated. A new favorite on the farm.

'Sherwood's Peach'

FORM: Informal Decorative

BLOOM SIZE: 10 to 12 in (25 to 30 cm)

The large, bronzy petals are backed by the softest purple haze, giving the flowers a dusty heirloom quality rarely seen in a dahlia. Flowers are on the more delicate side but are worth the extra effort for the color alone.

'Sierra Glow' ♥

FORM: Informal Decorative

BLOOM SIZE: 8 to 10 in (20 to 25 cm)

The peachy-salmon blooms on this long-time favorite are borne in abundance. Long stems nod slightly under the weight of the heavy flowers. They're a favorite autumn bouquet addition.

'Terracotta'

FORM: Semi-Cactus

BLOOM SIZE: 4 to 5 in (10 to 13 cm)

This variety has been a long-standing farm favorite. Blooms are a warm amber, reminiscent of butterscotch. Long, strong stems are perfect for arranging, and plants are extremely productive.

'Verrone's Socrates' ♥

FORM: Formal Decorative

BLOOM SIZE: Up to 4 in (10 cm)

This is one of our all-time favorite peach bloomers. Flowers are very weather resistant. It's great for weddings, and the extremely long stems lend themselves well to arranging.

# ORANGE

Filled with glowing sunset tones, this collection has
varying levels of saturation from clear bright orange,
marigold, and tangerine to intense shades of paprika,
burnt umber, and saffron.

'Amber Queen'

FORM: Pompon

BLOOM SIZE: 2 in (5 cm)

One of the earliest varieties to flower, this abundant bloomer is a real garden workhorse. Its petite, glowing bronze button-shaped blooms make great cuts and are a wonderful addition to bouquets. Good in pots.

'Babylon Bronze'

FORM: Informal Decorative

BLOOM SIZE: 8 to 10 in (20 to 25 cm)

This award winner features blooms that contrast against dark foliage. Not top heavy like other dinner plate varieties, its strong stems and upward-facing flowers make it great for cutting and arranging. This variety is perfect for hot, humid climates.

'Baron Katie'

FORM: Informal Decorative

BLOOM SIZE: 4 to 6 in (10 to 15 cm)

Vivid orange blooms have petals that are slightly fluted and twisted, giving them a feathery quality. Stems are long and strong.

'Beatrice'

FORM: Ball

BLOOM SIZE: Up to 4 in (10 cm)

It is hard to find a really great orange dahlia with vibrant blooms that holds up in adverse weather, and this fits the bill. Great for cutting, the flowers bloom abundantly all season long on tall stems.

'Bed Head'

FORM: Incurved Cactus

BLOOM SIZE: 4 to 6 in (10 to 15 cm)

This crazy novelty is aptly named. The tall plants are loaded with an abundance of wild-looking flowers that have incurved, tangerine petals. They make wonderful cut flowers and are always a favorite with everyone who visits our garden.

'Ben Huston'

FORM: Informal Decorative

BLOOM SIZE: 10+ in (25+ cm)

Flowers start out as a warm apricot with darker brush marks, then the centers fade to yellow. Long, strong stems make this an excellent addition to fall bouquets.

### 'Bloomquist Jean'

FORM: Informal Decorative

BLOOM SIZE: 6 to 8 in (15 to 20 cm)

Tall, vigorous plants are loaded with coppery flowers that have pointed, reflexed petals and remind us of the Muppets. They are iridescent and shimmer in the garden.

### 'Bloomquist Pumpkin'

FORM: Stellar

BLOOM SIZE: 6 to 8 in (15 to 20 cm)

Flowers have burnt sienna outer petals and glowing yellow centers. This fun variety looks windblown and relaxed.

### 'Bright Star'

FORM: Incurved Cactus

BLOOM SIZE: 4 to 6 in (10 to 15 cm)

This variety has such a fitting name. Cheerful pumpkin-orange flowers with pointed petals sit atop long, strong stems. These productive bloomers are standouts in flower arrangements.

### 'Brown Sugar'

FORM: Ball

BLOOM SIZE: Up to 4 in (10 cm)

Rusty red-orange flowers have lower petals that soften to warm terra-cotta as blooms age. Ball-shaped flowers are weather resistant and great for straight bunches.

### 'Camano Buz' ♥

FORM: Formal Decorative

BLOOM SIZE: Up to 4 in (10 cm)

A new favorite and one of the best orange varieties we've ever grown. Plants are smothered in warm-hued blooms all season long. Strong stems are great for cutting and the perfect size for bouquets.

### 'Catalina' ♥

FORM: Ball

BLOOM SIZE: Up to 4 in (10 cm)

Beautiful cantaloupe-colored flowers have orange undertones, dark green foliage, and long, strong stems. It's a real winner for straight bunches.

'Cornel Bronze' ♥

FORM: Ball

BLOOM SIZE: 4 in (10 cm)

Hands down, this is one of the best orange varieties on the market, with bronze petals and long, strong stems that are perfect for cutting. Flowers are weather resistant and last an extremely long time in the vase.

'Crazy 4 Vinnie' ♥

FORM: Formal Decorative

BLOOM SIZE: Up to 4 in (10 cm)

An excellent cut flower, this highly uniform grower has soft, clear pumpkin-orange blooms. Growth habit is very upright, and the upward-facing flowers look distinctive on their strong, dark stems.

'David Digweed'

FORM: Formal Decorative

BLOOM SIZE: 4 in (10 cm)

Flowers have darker centers and clear, light green eyes. This variety produces abundantly and is great for straight bunches.

'Eileen C'

FORM: Anemone

BLOOM SIZE: 3½ in (9 cm)

A favorite from our trials, this variety has brilliant persimmon and tangerine flowers. Large, round outer petals encircle golden-orange, fluffy centers, giving the long-lasting blooms a tropical feel.

'Ginger Willo' ♥

FORM: Pompon

BLOOM SIZE: 2 in (5 cm)

Pairing beautifully with just about anything, this variety is a lovely blend of warm tangerine and burnt orange. The blooms top super long, thin stems. Extremely productive and perfect for boutonnieres and bouquets—it's no wonder floral designers go crazy over this little guy.

'Hy Patti' ♥

FORM: Formal Decorative

BLOOM SIZE: 6 to 8 in (15 to 20 cm)

One of the best orange varieties we've ever grown, this stunning coppery gem produces a steady abundance of blooms all season long. The medium-size flowers have perfectly placed petals that curve gently inward.

### 'Hy Suntan' ♥

**FORM:** Ball

**BLOOM SIZE:** 3 to 4 in (8 to 10 cm)

The medium, coppery, ball-shaped flowers of this showstopping variety are a farm favorite. Plants are on the shorter side, but if pinched early they will produce a bumper crop of long, strong stems that are excellent for cutting. It's a must-grow for arranging.

### 'Irish Glow'

**FORM:** Pompon

**BLOOM SIZE:** Up to 2 in (5 cm)

This darling orange-raspberry variety is smothered in miniature round blooms all season long. While plants are on the petite side, flowers ride atop long, strong stems, making them ideal for cutting and flower arranging.

### 'Jomanda' ♥

**FORM:** Formal Decorative

**BLOOM SIZE:** 4 to 6 in (10 to 15 cm)

One of the most productive varieties we grow, its blooms are a rusty orange with contrasting dark stems. Flowers are both beautiful and long lasting, making them a favorite with market customers.

### 'Jowey Linda'

**FORM:** Formal Decorative

**BLOOM SIZE:** 3 to 4 in (8 to 10 cm)

Glowing in the garden, these pumpkin-colored beauties are both hardy and extremely weather resistant. A productive bloomer, this is a great all-around variety.

### 'Lakeview Lucky'

**FORM:** Formal Decorative

**BLOOM SIZE:** 4 to 6 in (10 to 15 cm)

Flowers in a mix of peach and coral tones are borne in sprays, and one plant can have flower heads in multiple colors. Its great range of tones makes it versatile for flower arranging.

### 'Lyn's Brooke'

**FORM:** Formal Decorative

**BLOOM SIZE:** 4 to 6 in (10 to 15 cm)

This dark peach beauty was a standout in our trial patch. The large plants are loaded with long-stemmed flowers all season long, making it a great variety for cutting.

'Maarn' (also sold as 'Sylvia')

FORM: Mini Ball

BLOOM SIZE: 2 to 3½ in (5 to 9 cm)

These bright, cheerful flowers are a huge hit with market customers. The extremely productive plants churn out an abundance of blooms all season long.

'Parkland Glory'

FORM: Informal Decorative

BLOOM SIZE: 8 to 10 in (20 to 25 cm)

Embracing the color of autumn, these red-orange blooms have gold edging around the tips. Plants are on the shorter side, and strong, dark stems hold the upward-facing flowers.

'Pooh'

FORM: Collarette

BLOOM SIZE: 2 to 4 in (5 to 10 cm)

The tall plants churn out loads of bright red-orange, single-petaled flowers in abundance. Unlike most collarette varieties, this novelty holds up well in the vase, making it an excellent cut flower.

'Punkin Spice'

FORM: Informal Decorative

BLOOM SIZE: 6 to 7 in (15 to 18 cm)

This beauty boasts glowing orange blooms, some with raspberry undertones and others with hints of gold and red. Petals are laciniated at the tip, giving the flowers an almost fuzzy quality. Their lush growth requires extra support.

'Rose Toscano' ♥

FORM: Formal Decorative

BLOOM SIZE: 3 to 4 in (8 to 10 cm)

Soft apricot blooms ride atop long, strong stems, making them ideal for flower arranging. It's especially popular with wedding and event designers.

'Rossendale Peach'

FORM: Formal Decorative

BLOOM SIZE: 4 to 6 in (10 to 15 cm)

Tall, strong stems carry blooms in clear orange. This cheerful variety has a beautiful flower shape that's stunning in arrangements.

**'Sandia Bill J'**

FORM: Anemone

BLOOM SIZE: 4 to 6 in (10 to 15 cm)

Large star-shaped flowers have outer petals of raspberry streaked with melon along with fluffy apricot centers. Stems are long and wiry.

**'Sylvia Craig Hunter'**

FORM: Formal Decorative

BLOOM SIZE: 6 to 8 in (15 to 20 cm)

These rich pumpkin-orange flowers are slightly nodding. Dark, long, strong stems hold blooms on this vigorous, upright grower.

**'Valley Rust Bucket'** ♥

FORM: Mini Ball

BLOOM SIZE: 2 to 3½ in (5 to 9 cm)

This garden workhorse was lovingly bred by Leone and David Smith. It has deep, rust-colored blooms that are a favorite for everyone who visits the garden. Productive plants produce tall, weather-proof, long-lasting stems all season long.

**'Verrone's 14-30'** ♥

FORM: Laciniated

BLOOM SIZE: 4 to 6 in (10 to 15 cm)

Glowing orange sherbet flowers have snipped petals that surround a pronounced glowing green center. Strong stems hold flowers upright, and plants are very productive. It's one of my all-time favorite orange varieties.

**'Verrone's Sandra J'**

FORM: Formal Decorative

BLOOM SIZE: 4 to 6 in (10 to 15 cm)

In shades of rust, raspberry, and gold, flowers look like an autumn sunset. Petal tips are touched with raspberry, adding visual interest. Towering plants have strong stems.

**'Wyn's Moonlight Sonata'**

FORM: Informal Decorative

BLOOM SIZE: 6 to 8 in (15 to 20 cm)

This is one of the most unique varieties we've grown. Its giant wavy petals are the most glorious blend of salmon-pink, coral, and tangerine. No two flowers are the same.

# CORAL

Warm tropical tones including papaya, persimmon, watermelon, pink grapefruit, and warm salmon-pink fill this category. These colors are extremely versatile and mix beautifully with most others.

### 'Aljo'

**FORM:** Semi-Cactus

**BLOOM SIZE:** 6 to 8 in (15 to 20 cm)

Stunning salmon blooms have petals that are folded and dusted with yellow. Flowers are sturdy for their shape and have long, strong, chocolate-brown stems.

### 'Askwith Minnie' ♥

**FORM:** Formal Decorative

**BLOOM SIZE:** 6 to 8 in (15 to 20 cm)

One of the best corals we have seen, these flowers' versatile color has a tropical feel and stands out against super long, strong dark stems. A vigorous grower with a marvelous shape.

### 'Bernadette Castro'

**FORM:** Formal Decorative

**BLOOM SIZE:** 4 in (10 cm)

Glowing soft coral flowers with peach undertones grow on productive plants that are on the shorter side.

### 'Bloomquist Beth'

**FORM:** Incurved Cactus

**BLOOM SIZE:** 6 to 8 in (15 to 20 cm)

This fun variety has quilled apricot blooms with coral tips and edging and yellow throats. An unusual addition for both the garden and the vase.

### 'Carmen Fiesta'

**FORM:** Formal Decorative

**BLOOM SIZE:** 6 to 8 in (15 to 20 cm)

This strong grower churns out so many flowers. It's perfectly named—blooms in soft butter yellow have cherry-red freckles and streaks that are fun and festive.

### 'Cecil' ♥

**FORM:** Formal Decorative

**BLOOM SIZE:** 2 to 3½ in (5 to 9 cm)

A favorite with designers, these blooms have long stems that show off their striking display of mixed coral, peach, and apricot hues.

'Crazy 4 Don'

**FORM:** Formal Decorative

**BLOOM SIZE:** 4 to 6 in (10 to 15 cm)

The gorgeous smoky-peach flowers of this variety sit atop strong stems with glossy green foliage. Its exquisite coloring is versatile for flower arranging.

'Fur Elise'

**FORM:** Formal Decorative

**BLOOM SIZE:** 3 to 5 in (8 to 13 cm)

Glowing coral-orange flowers have raspberry-backed petals. In addition, petal tips are pointed and slightly reflexed, providing extra dimension.

'Hillcrest Kismet' ♥

**FORM:** Formal Decorative

**BLOOM SIZE:** 5 to 6 in (13 to 15 cm)

Great in dramatic arrangements, these blooms are a warm salmon orange. Slightly nodding flowers rise above long, strong stems. A must-have if you make a lot of large-scale arrangements.

'Ice Tea' ♥

**FORM:** Formal Decorative

**BLOOM SIZE:** 3 to 4 in (8 to 10 cm)

This gem was a standout favorite in our trial patch. Compact plants are great in borders and produce an abundance of long-stemmed, raspberry-peach blooms that are ideal for flower arranging.

'Intrigue'

**FORM:** Formal Decorative

**BLOOM SIZE:** 4 in (10 cm)

This is one of the first dahlia varieties to flower each season. Blooms start out as a brilliant coral that fades to rich raspberry. It's prolific, long lasting, and loved by market customers.

'Janice'

**FORM:** Waterlily

**BLOOM SIZE:** 4 to 5 in (10 to 13 cm)

Warm salmon-pink flowers are upward facing, with outer petals that fade to cream. Long, strong stems are great for hand-tied bouquets, and plants are very productive.

### 'Jitterbug'

**FORM:** Formal Decorative

**BLOOM SIZE:** Up to 4 in (10 cm)

This super adorable flower has warm bubblegum-pink petals with yellow undertones surrounding a lavender center. These gems are on the shorter side.

### 'Kenora Lisa' ♥

**FORM:** Formal Decorative

**BLOOM SIZE:** 6 to 8 in (15 to 20 cm)

This coral-salmon beauty features blooms that are extremely popular with floral designers. Great in wedding work, plants produce a steady stream of strong-stemmed, beautiful blooms all season long.

### 'Lakeview Storm' ♥

**FORM:** Informal Decorative

**BLOOM SIZE:** 4 to 6 in (10 to 15 cm)

Flowers in autumn sunset tones of raspberry to smoky orange have petals that twist, giving blooms a shaggy quality. Tall plants produce dark, wiry stems that are great for arranging.

### 'Lynette'

**FORM:** Semi-Cactus

**BLOOM SIZE:** 4 to 6 in (10 to 15 cm)

Raspberry, upward-facing blooms have lavender dusting and stripes. Petals are rolled back and pointed on long, dark branching stems.

### 'NTAC Mia Li'

**FORM:** Formal Decorative

**BLOOM SIZE:** 3 to 5 in (8 to 13 cm)

These productive, hardy plants produce unusual glowing coral flowers with apricot and gold undertones. Upward-facing blooms carried on long, strong stems are ideal for mixed and straight bunches.

### 'Pam Howden'

**FORM:** Waterlily

**BLOOM SIZE:** 4 to 6 in (10 to 15 cm)

A perfect waterlily shape, this variety has elegant flowers held upright on long, strong stems. The flat, saucer-shaped blooms have a yellow base with peach tips accented by raspberry backs.

### 'Penhill Watermelon'

**FORM:** Informal Decorative

**BLOOM SIZE:** Over 10 in (25 cm)

One of the most beautiful dinner plate varieties we've ever grown, this giant fluffy gem is a unique mix of peach, lavender, and the tiniest hint of yellow. Flowers are borne in abundance, and everyone who sees it falls in love.

### 'Princess Elisabeth'

**FORM:** Formal Decorative

**BLOOM SIZE:** Up to 4 in (10 cm)

Upward-facing blooms are an unusual smoky raspberry with metallic gold dusting. They are a great size for arranging. Pick flowers when they are on the younger side, since the petals have a tendency to drop on the mature flowers.

### 'Robin Hood'

**FORM:** Ball

**BLOOM SIZE:** 4 in (10 cm)

Large flowers are a stunning mix of coral, peach, and apricot. Well loved by both market and floral design customers, these plants are very productive.

### 'Sebastian'

**FORM:** Formal Decorative

**BLOOM SIZE:** 3 in (8 cm)

Flowers that practically glow are a warm melon color with a peach-lavender center. Dark foliage covers extra-productive plants that are on the shorter side.

### 'September Morn'

**FORM:** Formal Decorative

**BLOOM SIZE:** 5 in (13 cm)

One of the hardest-working varieties we've encountered that consistently outperforms other plants. Blooms are a mix of raspberry, gold, and peach, and as nights cool in autumn, flowers darken in color. Upward-facing flowers are great for bunches.

### 'Showcase Decadent'

**FORM:** Formal Decorative

**BLOOM SIZE:** 3 to 6 in (8 to 15 cm)

This color is so lovely and versatile. Warm salmon-coral blooms are held on tall, dark stems, contrasting beautifully. A wonderful variety for arranging.

## 'Snoho Doris' ♥

**FORM:** Ball

**BLOOM SIZE:** 4 to 5 in (10 to 13 cm)

Award-winning and one of our top five favorites, this vigorous, upright grower has long, strong stems and virtually weatherproof flowers. A favorite with designers and market customers, blooms are a beautiful mix of coral, peach, and apricot.

## 'Snoho JoJo' ♥

**FORM:** Ball

**BLOOM SIZE:** Over 3½ in (9 cm)

This beautiful bronze bloomer is as productive as it is pretty. Plants churn out an abundance of long-stemmed, cheerful flowers all season.

## 'Summer Beauty' ♥

**FORM:** Informal Decorative

**BLOOM SIZE:** 8 to 10 in (20 to 25 cm)

One of the most beautiful coral varieties we've grown, its flowers glow with pointed gold-dipped tips. The petals look as if they've been brushed with glitter.

## 'Sun Spot'

**FORM:** Ball

**BLOOM SIZE:** Over 3½ in (9 cm)

Salmon-peach blooms with warm yellow centers grow on long, strong stems. This variety with a fitting name is great for straight bunches.

## 'Wildwood Marie'

**FORM:** Waterlily

**BLOOM SIZE:** 4 to 5 in (10 to 13 cm)

Long-stemmed blossoms have peachy-coral petals with soft yellow undertones. Upward-facing blooms and delightful coloring are great for bouquets.

## 'Yvonne'

**FORM:** Waterlily

**BLOOM SIZE:** 4 in (10 cm)

Glowing, upward-facing coral blooms have a tropical feel reminiscent of a sunrise. Flowers are held atop long wiry stems and are perfect for hand-tied bouquets.

# RASPBERRY

This group possesses vibrant blooms with a strangely haunting quality and oftentimes an antiqued appearance. Colors come from a rich palette of delicious fruity shades, including sangria, a smoky peach with a lavender dusting, and juicy raspberry.

## 'AC Thornbird'

**FORM:** Laciniated

**BLOOM SIZE:** 6 to 8 in (15 to 20 cm)

This fun variety produces a bumper crop of long-stemmed flowers with split petal tips that look like they've been snipped with pinking shears.

## 'All That Jazz' ♥

**FORM:** Formal Decorative

**BLOOM SIZE:** 4 to 6 in (10 to 15 cm)

This favorite's rare coloring reminds us of raspberry lemonade and seems to change depending on the weather; petal tips are slightly lighter, providing extra dimension. Strong, tall plants are very productive.

## 'American Dawn'

**FORM:** Formal Decorative

**BLOOM SIZE:** 6 to 8 in (15 to 20 cm)

Truly one of a kind, this unique bloomer has the loveliest purple center and reverse petals. The long, strong stems, dark foliage, and big blooms make it a must-have for flower arrangers.

## 'Andy's Legacy'

**FORM:** Informal Decorative

**BLOOM SIZE:** 6 to 8 in (15 to 20 cm)

Layered and folded petals give this smoky-peach treasure a ruffled look. Flowers are held on long dark stems and are great for wedding work.

## 'Bacardi'

**FORM:** Formal Decorative

**BLOOM SIZE:** 4 to 6 in (10 to 15 cm)

One of the most beautifully colored varieties we grow, this muddy-rose bloomer has dark raspberry tips and centers. Harvest when buds are one-third of the way open, since the delicate petals are more prone to weather damage than most.

## 'Barbarry Esquire'

**FORM:** Formal Decorative

**BLOOM SIZE:** Up to 4 in (10 cm)

This gem has magenta blooms with maroon-purple centers. A glowing gold undertone and frosted petal tips give them an iridescent quality.

**'Barbarry Imperial'** ♥

**FORM:** Formal Decorative

**BLOOM SIZE:** 4 to 6 in (10 to 15 cm)

With glowing watermelon blooms, there's no other color like this. Long, strong stems grow on vigorous plants.

**'Belle of Barmera'**

**FORM:** Informal Decorative

**BLOOM SIZE:** 10+ in (25 cm)

The peach-centered, coral-raspberry blooms of this giant beauty are a magnificent sight to behold. The towering plants are loaded with flowers on long, strong stems, perfect for large-scale flower arrangements.

**'BJ's Rival'** ♥

**FORM:** Anemone

**BLOOM SIZE:** 3 in (8 cm)

This novelty displays one of the best color combinations and flower shapes we've seen. Plants are smothered in fluffy, two-toned raspberry-and-gold blooms that remind us of echinaceas. Though the flowers have a delicate appearance, they hold up well in weather.

**'Blue Peter'**

**FORM:** Incurved Cactus

**BLOOM SIZE:** 10+ in (25 cm)

Slightly nodding, spidery flowers have swirled, twirled petals in a warm apricot tone with a lavender-blush cast on the backs of petals.

**'Bryn Terfel'**

**FORM:** Informal Decorative

**BLOOM SIZE:** 10+ in (25 cm)

Massive coral-raspberry blooms sit atop long, strong stems. Petals twist and fade, giving them a unique multidimensional appearance.

**'Caitlin's Joy'**

**FORM:** Ball

**BLOOM SIZE:** Over 3½ in (9 cm)

Metallic raspberry blooms grow on strong stems. These sturdy, weather-resistant flowers are very productive and great for straight bunches.

### 'Camano Mordor'

**FORM:** Ball

**BLOOM SIZE:** 4 in (10 cm)

The base of this full, densely packed flower is yellow, and petal tips are raspberry-coral. Blooms include a striking range of sunset hues and are great for cutting.

### 'Camano Mystery'

**FORM:** Formal Decorative

**BLOOM SIZE:** 4 to 6 in (10 to 15 cm)

Long, strong stems are topped with glowing flowers. Pointed petals have dustings of gold that give them a fun tropical feel.

### 'Chewy'

**FORM:** Formal Decorative

**BLOOM SIZE:** 3 in (8 cm)

Medium-size bushy plants have flowers that are peachy buff with the lightest lavender dusting, giving them a bronzed look.

### 'Daisy Duke' ♥

**FORM:** Formal Decorative

**BLOOM SIZE:** Up to 4 in (10 cm)

After years of searching for this color, we came across these blooms in pink-salmon-coral with a mauve center. Short, stocky plants require pinching and hard cutting to produce long stems, but the distinctive color makes it worth the extra effort. Good in pots.

### 'Ferncliff Carefree'

**FORM:** Mini Ball

**BLOOM SIZE:** 2½ to 3 in (6 to 8 cm)

Tall, branching plants are loaded with a bumper crop of salmon-pink blooms. Long stems are excellent for cutting.

### 'Ferncliff Rusty' ♥

**FORM:** Ball

**BLOOM SIZE:** Over 3½ in (9 cm)

Deep watermelon flowers have a moody quality. Tall stems are great for straight bunches. Plants produce a bumper crop of blooms.

### 'Foxy Lady'

**FORM:** Formal Decorative

**BLOOM SIZE:** Up to 4 in (10 cm)

Flowers are a dusty mauve with a soft buttercream background. The reverse of the petals is darker, giving them a very unusual look. One of the first varieties to bloom every summer, this productive variety is a standout.

### 'Hy Zizzle'

**FORM:** Novelty Open

**BLOOM SIZE:** 3 in (8 cm)

This super fun novelty has pointed magenta outer petals with a secondary layer of shorter fringed petals encircling a quilled golden terra-cotta center. Long, branching stems arise from textural ferny foliage.

### 'Jabberbox' ♥

**FORM:** Formal Decorative

**BLOOM SIZE:** Up to 4 in (10 cm)

Tall, productive plants are loaded with a bumper crop of blooms in soft peach with raspberry-coral streaking that have tons of dimension. Stems are long and strong.

### 'Janet Allison'

**FORM:** Informal Decorative

**BLOOM SIZE:** 6 in (15 cm)

This beauty is a rare treasure. Blooms are an unusual mix of raspberry and gold and change substantially throughout the season.

### 'Jive'

**FORM:** Anemone

**BLOOM SIZE:** 3½ in (9 cm)

Large star-shaped raspberry blooms have fluffy centers dusted in gold and textural ferny foliage. This fun novelty has stunning color.

### 'Jowey Frambo'

**FORM:** Mini Ball

**BLOOM SIZE:** 2 to 3½ in (5 to 9 cm)

This treasure was a standout in a past dahlia trial. The bright raspberry-pink blooms are borne in abundance all season long and ride atop tall strong stems.

## 'Jowey Winnie' ♥

**FORM:** Ball

**BLOOM SIZE:** Over 3½ in (9 cm)

The vigorous, hardy plants churn out armloads of rose flowers brushed with lavender all season long. Their strong stems make them ideal for cutting. A must-have color for flower arrangers and wedding florists.

## 'KA's Rosie Jo' ♥

**FORM:** Formal Decorative

**BLOOM SIZE:** Up to 4 in (10 cm)

This fantastic introduction from Santa Cruz Dahlias boasts incredible smoky raspberry flowers on strong, dark stems. Its rare and beautiful coloring puts it high on our list.

## 'Labyrinth' ♥

**FORM:** Informal Decorative

**BLOOM SIZE:** 8 to 10 in (20 to 25 cm)

Trying to put the beauty of this magnificent variety into words feels almost impossible. The vigorous, dark-leaved plants produce a staggering number of large peachy-raspberry flowers that are like nothing else we've ever grown.

## 'Mission Gypsy'

**FORM:** Formal Decorative

**BLOOM SIZE:** Up to 4 in (10 cm)

The knee-high plants produce rounded blooms that are a stunning rich raspberry with a gentle lavender dusting that gives it a mysterious quality.

## 'Ms Prissy'

**FORM:** Stellar

**BLOOM SIZE:** 4 to 6 in (10 to 15 cm)

Shorter plants produce juicy watermelon-colored flowers. Petals on this beauty have a wavy quality, and blooms are a great size for flower arranging.

## 'Mystique' ♥

**FORM:** Formal Decorative

**BLOOM SIZE:** 4 in (10 cm)

Plants produce an abundance of long-stemmed, dusty-rose flowers all season long. The outside edges of the petals gently fade, creating a smoky effect—truly a sight to behold.

### 'Omega'

**FORM:** Laciniated

**BLOOM SIZE:** 10 + in (25 cm)

Large plants and long, strong stems hold up smoky coral blooms that have an iridescent quality. This variety is a standout for flower arranging because of its versatile coloring.

### 'Penhill Dark Monarch'

**FORM:** Informal Decorative

**BLOOM SIZE:** 10 to 12 in (25 to 30 cm)

One of the most striking large-flowered varieties we grow, the smoky-plum blooms of this moody giant are loved by all who visit our garden. Plants produce an abundance of huge, ruffled flowers on long stems.

### 'Pipsqueak'

**FORM:** Collarette

**BLOOM SIZE:** 3 in (8 cm)

Glowing raspberry outer petals surround an iridescent, slightly lavender center ring. Plants are on the shorter side and are prolific producers. This cutie looks beautiful in the front of the border, and the flowers are great for arranging.

190

### 'Polka' ❤

**FORM:** Anemone

**BLOOM SIZE:** 4 to 6 in (10 to 15 cm)

This one-of-a-kind novelty features layers of creamy petals streaked with cranberry that encircle a large, fluffy golden center. No two flowers are the same, and there's nothing else like it on the market. Blooms are both weather resistant and long lasting.

### 'Salmon Runner'

**FORM:** Formal Decorative

**BLOOM SIZE:** 4 in (10 cm)

Medium-size plants produce sprays of the most beautiful smoky coral-raspberry blooms. Petal edges and tips are slightly lighter, giving it added dimension.

### 'Sandia Panama'

**FORM:** Anemone

**BLOOM SIZE:** 3½ in (9 cm)

This variety produces an abundance of petite blooms that seem to dance in the breeze above attractive ferny foliage. Outer petals are a rich cherry red, fading to ivory at the tips, and centers are gold and fuzzy.

### 'Tempest'

**FORM:** Formal Decorative

**BLOOM SIZE:** 5 in (13 cm)

Warm coral-raspberry flowers are dusted ever so slightly with lavender. This uniquely colored variety is a longtime favorite.

### 'Totally Tangerine'

**FORM:** Anemone

**BLOOM SIZE:** 3 to 4 in (8 to 10 cm)

This is a favorite with all who visit our garden. The unique tangerine flowers have a soft blush of rose pink on the back side of the petals, creating the most magical effect. Compact plants are great in pots.

### 'Vista Minnie'

**FORM:** Novelty Fully Double

**BLOOM SIZE:** 4½ in (11 cm)

This stunner has petals that are peach on the front and raspberry on the back. Petals roll and curl, giving it a two-toned effect that looks like a pinwheel.

### 'Waltzing Mathilda' ♥

**FORM:** Single

**BLOOM SIZE:** 4 in (10 cm)

The vivid coral-peach single flowers on this farm favorite are set against a sea of dark foliage and bloom abundantly, without being deadheaded, all season long. A marvelous variety for cutting and a spectacular addition to the flower border.

### 'Wannabee'

**FORM:** Anemone

**BLOOM SIZE:** 3 in (8 cm)

A striking addition to the garden, this variety displays vivid fuchsia-pink blooms with glowing, fluffy tangerine centers. Set against ferny foliage, these dark-stemmed beauties are great for arranging.

### 'Zundert Mystery Fox'

**FORM:** Mini Ball

**BLOOM SIZE:** 2 to 3½ in (5 to 9 cm)

Medium-size plants are extremely productive and give out a bumper crop of coral-orange blooms on long, strong stems. Flowers have a striking yellow-green eye.

# PINK

Feminine tones, including ballet slipper, dusty rose, frost pink, carnation, bubblegum, magenta, and fuchsia, fill this category. This color range is as sweet as can be and is quite popular for weddings.

### 'Alloway Candy' ♥

**FORM:** Stellar

**BLOOM SIZE:** 4 to 6 in (10 to 15 cm)

This darling baby-pink bloomer has won the hearts of all of our customers, especially wedding florists and brides. It flowers prolifically and makes a lovely addition to hand-tied bouquets.

### 'Bahama Mama'

**FORM:** Informal Decorative

**BLOOM SIZE:** 4 to 6 in (10 to 15 cm)

This vigorous grower has long, strong stems, and its blooms are peach-pink with lilac tips and a creamy center. This striking color range makes it a real standout.

### 'Bargaly Blush'

**FORM:** Formal Decorative

**BLOOM SIZE:** 5 to 8 in (13 to 20 cm)

Bubblegum-pink blooms on medium-size plants have long, strong stems. The sizable flowers are well suited for large-scale arrangements.

### 'Betty Anne' ♥

**FORM:** Pompon

**BLOOM SIZE:** Up to 2 in (5 cm)

Small plants are smothered in the most adorable lavender-pink blooms on long wiry stems. This is one of the most productive pompons we grow—each plant regularly produces 50 to 75 blooms per season.

### 'Bloomquist Isla'

**FORM:** Formal Decorative

**BLOOM SIZE:** 6 to 8 in (15 to 20 cm)

The medium-size creamy flowers have pink-tipped petal edges and a darker cranberry center. This lovely color is perfect for weddings.

### 'Bracken Rose' ♥

**FORM:** Formal Decorative

**BLOOM SIZE:** 4 in (10 cm)

This beauty is the most exquisite shade of dusty rose that we affectionately call "ballet slipper pink." Long stems are topped with slightly nodding flower heads that look incredible in arrangements.

### 'Camano Love'

**FORM:** Mini Ball

**BLOOM SIZE:** 2 to 3½ in (5 to 9 cm)

A vigorous grower, this variety produces a bumper crop of fuchsia flowers with a lavender cast that are sturdy and long lasting. Blooms have an unusual textural quality.

### 'Chilson's Pride'

**FORM:** Informal Decorative

**BLOOM SIZE:** 4 to 6 in (10 to 15 cm)

This darling features soft pink flowers with pale centers that smother the plants all season long. Petal tips are slightly fringed, providing extra dimension. Both color and size make it ideal for arranging.

### 'Chimacum Katie' ♥

**FORM:** Formal Decorative

**BLOOM SIZE:** 4 to 6 in (10 to 15 cm)

One of the very best varieties in this colorway, its glowing bright pink flowers top tall, strong stems. Ultra-productive blooms are uniform and excellent for cutting.

### 'Colorado Classic'

**FORM:** Informal Decorative

**BLOOM SIZE:** 4 to 6 in (10 to 15 cm)

This striking dahlia has pointed, wavy ivory petals that are edged and tipped with vivid candy pink tones. Flowers are held on long stems, and it's a stunner in the garden.

### 'Fancy Pants' ♥

**FORM:** Orchette

**BLOOM SIZE:** 3½ to 4 in (9 to 10 cm)

Tall, vigorous plants produce the most beautiful cerise-pink-and-white streaked blooms that fold and twirl. They look exactly like pinwheels, and bees love them.

### 'Gay Princess'

**FORM:** Informal Decorative

**BLOOM SIZE:** 4 to 6 in (10 to 15 cm)

Bubblegum-pink flowers have laciniated tips, giving the flowers a fuzzy appearance. This is a sturdy, vigorous upright grower, and flowers are held atop long stems.

### 'Gerrie Hoek'

**FORM:** Waterlily

**BLOOM SIZE:** 5 to 6 in (13 to 15 cm)

It's no wonder this old-fashioned favorite is still so widely grown. Plants are smothered in pretty, soft pink blooms, and the upward-facing flowers and strong stems make them ideal for cutting and arranging.

### 'Hollyhill Cotton Candy'

**FORM:** Incurved Cactus

**BLOOM SIZE:** 6 to 8 in (15 to 20 cm)

Large flowers look just like sea anemones. Upward-facing flowers have tubular petals. They're a fun addition to the garden and arrangements.

### 'Hollyhill Pinkie' ❤

**FORM:** Laciniated

**BLOOM SIZE:** 4 in (10 cm)

Tall plants have soft candy pink flowers that are edged in purple. Petals have laciniated tips that give a fluffy, feathered quality.

### 'Jan Van Schaffelaar'

**FORM:** Pompon

**BLOOM SIZE:** Up to 2 in (5 cm)

These vigorous plants are smothered in an abundance of bright pink flowers. This vivid bloomer is striking in the garden, and the blooms are the perfect size for bouquets.

### 'Miss Delilah'

**FORM:** Informal Decorative

**BLOOM SIZE:** 6 in (15 cm)

An orchid-pink base of petals blends to a buttercream center and dark eye. This is a medium-size plant and strong grower, and the blooms are an excellent choice for straight bunches.

### 'Miss Teagan'

**FORM:** Semi-Cactus

**BLOOM SIZE:** 4 to 6 in (10 to 15 cm)

Undoubtedly one of the best soft pink dahlias we've encountered, this variety's flowers are shell pink brushed with lavender and cream. These productive plants bloom on long stems.

### 'Otto's Thrill'

**FORM:** Informal Decorative

**BLOOM SIZE:** 8 to 12 in (20 to 30 cm)

This giant rosy-pink dinner plate is a crowd favorite, especially with florists. With long, strong stems and huge, shimmering blossoms, this deserves a spot in every cutting garden.

### 'Pinelands Princess'

**FORM:** Laciniated

**BLOOM SIZE:** 6 to 8 in (15 to 20 cm)

A strong grower and delightful flower, this variety is striking with its distinct shape and colors. Creamy petals dipped in fuchsia look like they have been clipped with pinking shears. It's fun and eye-catching in the garden.

### 'Pink Runner'

**FORM:** Formal Decorative

**BLOOM SIZE:** 2 to 4 in (5 to 10 cm)

This flower's lavender-pink petals with pale yellow undertones have an iridescent quality. These beauties are displayed on thick stems, and their upward-facing blooms are great for arranging.

### 'Pink Sylvia'

**FORM:** Ball

**BLOOM SIZE:** Over 3½ in (9 cm)

This variety has lush upright growth and striking dark stems. Fuchsia-pink flowers with frosted edges make for added dimension and interest.

### 'Rosy Wings'

**FORM:** Collarette

**BLOOM SIZE:** 3 to 4 in (8 to 10 cm)

Petite in stature, this bloomer still produces stems that are long enough for cutting, with flowers that look like cosmos. Petals are a lovely lavender-pink with slightly lighter petals ringing the golden center. Good in pots.

### 'Sandia Pouffe' ♥

**FORM:** Anemone

**BLOOM SIZE:** 3½ in (9 cm)

One of our favorite discoveries, this variety is loaded with ultra-full candy pink blooms. Flowers have a light dusting of gold at the tips, and plants are vigorous.

### 'Skipley Lois Jean' ♥

**FORM:** Ball

**BLOOM SIZE:** Over 3½ in (9 cm)

This fantastically productive beauty is one of the best bright pink varieties on the market. Its tall, strong stems and vivid, weatherproof blooms produce abundantly all season long.

### 'Tahoma April'

**FORM:** Ball

**BLOOM SIZE:** Over 3½ in (9 cm)

This garden workhorse churns out an abundance of long, strong-stemmed, dark pink flowers all season. It's extremely productive and great for cutting.

### 'Tahoma Little One'

**FORM:** Formal Decorative

**BLOOM SIZE:** 4 to 6 in (10 to 15 cm)

Creamy flowers are brushed with soft pink on the petal tips. This is a great variety for arranging and weddings.

### 'Tahoma Lucas'

**FORM:** Ball

**BLOOM SIZE:** Over 3½ in (9 cm)

Small plants are loaded with glowing flowers that look like cotton candy. Blooms have a multidimensional quality and are perfect for bouquets.

### 'Verrone's 14-64'

**FORM:** Stellar

**BLOOM SIZE:** 4 to 6 in (10 to 15 cm)

The medium-size glowing magenta-pink flowers have petals that point at the tip and reflex to reveal dense glowing yellow centers. It's extremely striking in the garden.

### 'Wyoming Wedding'

**FORM:** Informal Decorative

**BLOOM SIZE:** 4 to 6 in (10 to 15 cm)

Vigorous plants produce flowers on long, strong stems. Wavy pink petals give the blooms an almost fluffy appearance. A fun addition for the garden and arranging.

# PURPLE

This category boasts a wide array of tones and varying levels of saturation, including pale lavender, lilac, dusty mauve, plum, violet, and grape. These colors have a loyal following and oftentimes look best on their own.

### 'Chimacum Del Blomma'

**FORM:** Mini Ball

**BLOOM SIZE:** 2 to 3½ in (5 to 9 cm)

Likely the darkest purple dahlia we grow, its blooms are striking in the garden. Medium-size plants with rich grape flowers are dazzling and ideal for market sales.

### 'Clearview Lila'

**FORM:** Stellar

**BLOOM SIZE:** 4 to 6 in (10 to 15 cm)

Pale petals with pointed tips and a pearl-white cast form a domed shape that gives these blooms a multidimensional quality. Medium-size plants hold flowers on long, strong stems.

### 'Clifton Jordi' ♥

**FORM:** Formal Decorative

**BLOOM SIZE:** 4 to 6 in (10 to 15 cm)

Flowers start out a rosy pink and fade to cream with blush-purple tips that have a shimmering quality. Plants are on the shorter side.

### 'Crazy Cleere's' ♥

**FORM:** Mini Ball

**BLOOM SIZE:** 2 to 3½ in (5 to 9 cm)

This beauty is one of the most productive mini balls that we grow on the farm. Plants are loaded with adorable pink-cream and lavender flowers. A great color combo and nice size for bouquets.

### 'Dad's Favorite'

**FORM:** Anemone

**BLOOM SIZE:** 5 in (13 cm)

These purple flowers are unusual and eye-catching with their lovely dark violet petals and fuzzy, gold-flecked centers. Plants produce long stems prolifically all season.

### 'Dancin' Queen'

**FORM:** Stellar

**BLOOM SIZE:** 7 in (18 cm)

Soft lilac blooms have white tips and paler petals toward the center that give flowers a frosted quality. Flowers are upward facing with reflexed petals.

### 'Fluffles'

**FORM:** Formal Decorative

**BLOOM SIZE:** 5 in (13 cm)

These eye-catching beauties look like blackberry cheesecake. They have unusual soft bluish-purple flowers with creamy white centers set against dark green foliage. Plants are very productive.

### 'Frank Holmes'

**FORM:** Pompon

**BLOOM SIZE:** 1½ to 2 in (4 to 5 cm)

One of the sweetest little dahlias in our patch, this variety produces armloads of vivid violet flowers with a darker purple wash. This variety flowers on long, strong stems all season long.

### 'Genova' ♥

**FORM:** Mini Ball

**BLOOM SIZE:** 2 to 3½ in (5 to 9 cm)

The vigorous but manageable plants produce a bumper crop of the most beautiful, soft lavender, ball-shaped flowers that have a distinctive dark eye. Tall, strong stems make them excellent cut flowers.

200

### 'Hilltop Glo'

**FORM:** Formal Decorative

**BLOOM SIZE:** 4 to 6 in (10 to 15 cm)

A beautiful color combination, blooms are pink with purple edging and tips. Though plants are on the short side, the flowers are large and held on long, strong stems.

### 'Hollyhill Liz'

**FORM:** Mini Ball

**BLOOM SIZE:** 2 to 3½ in (5 to 9 cm)

This sturdy, hardy plant is a great addition to the dahlia patch. Petite, plum blooms have a rich dark grape dusting, and flowers are held atop strong, dark stems. A vigorous grower, this variety is a fantastic choice for bouquets.

### 'Koko Puff' ♥

**FORM:** Pompon

**BLOOM SIZE:** 2 in (5 cm)

With blooms in the most exquisite smoky-mauve hue, this variety blends beautifully with many different color palettes. It flowers steadily all season long.

### 'Leila Savanna Rose'

FORM: Semi-Cactus

BLOOM SIZE: 5 in (13 cm)

Medium-size plants are vigorous and upright. Unusual lavender flowers have grape backing and streaking and are borne in sprays.

### 'Mambo' ♥

FORM: Anemone

BLOOM SIZE: 4 in (10 cm)

This is one of the most fun varieties we've grown. Plants are smothered in a cloud of fluffy blooms with outer petals in warm violet and centers that are a pouf of cranberry and cream.

### 'Mary's Jomanda' ♥

FORM: Ball

BLOOM SIZE: Over 3½ in (9 cm)

The stunning blooms of this glowing magenta variety are produced in abundance all season. This is a longtime favorite on the farm.

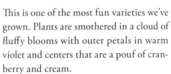

### 'Megan Dean'

FORM: Ball

BLOOM SIZE: 2 to 3½ in (5 to 9 cm)

Tall, extra-productive plants produce blush-lavender medium-size flowers that darken in the center. This beautiful soft color is great for bouquets.

### 'Mexico'

FORM: Anemone

BLOOM SIZE: 4 to 5 in (10 to 13 cm)

This lovely novelty has orchid-pink petals with glowing yellow and raspberry centers. Long, strong stems are smothered in blooms. A fun addition to the garden.

### 'Mi Wong'

FORM: Pompon

BLOOM SIZE: Up to 2 in (5 cm)

This adorable variety produces the prettiest rounded pale lavender-pink blooms with a lighter base. Great for wedding work. One of the smallest and sweetest varieties we grow on the farm.

'Nellie's Rose'

FORM: Pompon

BLOOM SIZE: Up to 2 in (5 cm)

Tiny grape-colored flowers have a honeycomb shape. Plants are on the smaller side and have a branching habit. Worth growing for the color alone.

'Nijinsky' ♥

FORM: Ball

BLOOM SIZE: Over 3½ in (9 cm)

So many qualities make this one of the best purple balls you can have. A strong grower, plants are extremely productive, churning out loads of smoky purple flowers.

'Puff-N-Stuff' ♥

FORM: Novelty Fully Double

BLOOM SIZE: 3½ in (9 cm)

Reminding us of an echinacea, this variety has warm red-purple outer petals that are flecked and streaked with color. Strongly reflexed, they surround an ultra-ruffled, darker center.

'Que Sera'

FORM: Anemone

BLOOM SIZE: 5 to 6 in (13 to 15 cm)

One of the most interesting in the patch, this upward-facing dahlia has white-and-purple petals surrounding fluffy golden centers. Stocky plants are covered in ferny foliage.

'RC Diane Brazil'

FORM: Stellar

BLOOM SIZE: 5 in (13 cm)

This strong grower is smothered in exceptional flowers. Grape and white streaks mark the backs of soft lavender reflexed petals, creating a whirling textural quality.

'Salish Going Dark' ♥

FORM: Formal Decorative

BLOOM SIZE: 4½ in (11 cm)

Washington grower Naomi "Noni" Morrison bred this long-lasting cut flower. Blooms are smoky-lavender with dark wine-purple outer petals that have a gunmetal cast. The ferny foliage adds even more interest and dimension.

### 'Senior's Hope'

**FORM:** Informal Decorative

**BLOOM SIZE:** 5 in (13 cm)

Flower petals are silvery mauve with rich plum backs and streaking along with dark centers. Plants are on the smaller side and have dark foliage, but the color is otherworldly.

### 'Shiloh Noelle'

**FORM:** Informal Decorative

**BLOOM SIZE:** 8 to 10 in (20 to 25 cm)

An exquisite dinner plate variety, this pale lavender treasure is a must-have for wedding work and large-scale arrangements.

### 'Skipley Spot of Gold' ♥

**FORM:** Formal Decorative

**BLOOM SIZE:** Up to 4 in (10 cm)

The rare and unusual coloring of this variety has quickly earned it a spot on our must-grow list for floral designers. Each rosy-lavender petal is tipped with a perfectly delicate spot of gold, creating a magical effect.

### 'Snoho Sonia'

**FORM:** Ball

**BLOOM SIZE:** Over 3½ in (9 cm)

These large, bushy plants are smothered in lavender-pink blooms. For a purple variety it's extremely soft and romantic.

### 'Tahoma Stellar Feller' ♥

**FORM:** Stellar

**BLOOM SIZE:** 5 to 6 in (13 to 15 cm)

Glowing snow-white flowers are dusted in the softest lavender blush with slightly darker centers. These tall plants have long stems perfect for arranging and wedding work.

### 'Take Off'

**FORM:** Anemone

**BLOOM SIZE:** 3 to 4 in (8 to 10 cm)

Soft, orchid-toned petals encircle a lighter, fluffy center on this special variety—no two flowers are the same. Blooms are accented by dark stems and ferny foliage, and the tall garden plants could be mistaken for peonies.

### 'Twilite'

**FORM:** Anemone

**BLOOM SIZE:** 4 in (10 cm)

This low-growing gem looks like a giant, double-flowered echinacea. Soft lavender-pink outer petals are reflexed back, revealing a cranberry and gold center.

### 'Vassio Meggos'

**FORM:** Informal Decorative

**BLOOM SIZE:** 8 to 10 in (20 to 25 cm)

This beautiful pale plum bloomer has long stems, huge flowers, and the ability to capture the attention of everyone who sees it.

### 'Vista Pinky'

**FORM:** Stellar

**BLOOM SIZE:** 5 in (13 cm)

With a misleading name, smoky metallic grape-colored flowers have a haunting antique quality. Back petals reflex, revealing a darker center. This variety is worth growing for the color alone.

204

### 'Willowfield Mathew' ♥

**FORM:** Formal Decorative

**BLOOM SIZE:** 3 in (8 cm)

One of the best lilac-rose varieties we've grown, this tall, ultra-productive beauty produces an abundance of long, strong stems and blooms all season long. One of the earliest varieties to flower on our farm.

### 'Wyn's Mauve Mist'

**FORM:** Semi-Cactus

**BLOOM SIZE:** 6 to 8 in (15 to 20 cm)

This vigorous grower has blooms that are a versatile soft violet shade with a gray wash. This muddy, muted effect lends itself to arranging, and dark stems give it a multidimensional quality.

### 'Zippity Do Da'

**FORM:** Pompon

**BLOOM SIZE:** Up to 2 in (5 cm)

Candy pink flowers have a darker purple eye. Vigorous plants with glossy green foliage are extra productive and absolutely adorable.

# RED

Striking and powerful in the garden and the vase,
this intense collection of colors includes ruby,
cardinal, scarlet, tomato soup, poppy, and even some
that are reminiscent of striped peppermint candies.

**'AC Rooster'** ❤

**FORM:** Stellar

**BLOOM SIZE:** 3 to 4 in (8 to 10 cm)

Fun tomato soup–red flowers have pointed petal tips that are swept back, giving it a feather-like quality. It's one of my favorite reds.

**'Alauna Pochette Surprise'**

**FORM:** Cactus

**BLOOM SIZE:** 4 to 6 in (10 to 15 cm)

Glowing coral-red flowers with white frosted tips remind us of pinwheels. This garden standout is a sight to behold.

**'Asahi Chohji'**

**FORM:** Anemone

**BLOOM SIZE:** 3 to 4 in (8 to 10 cm)

This is a really fun, cheerful novelty. Despite their delicate appearance, the stems are quite strong and look as if they are topped with striped peppermint candy.

**'Barberpole'**

**FORM:** Anemone

**BLOOM SIZE:** 4 to 5 in (10 to 13 cm)

Petals are a mix of red and white, with fluffy white-and-gold centers. This striking novelty has an old-fashioned candy-stripe quality.

**'Bloomquist Joel'**

**FORM:** Laciniated

**BLOOM SIZE:** 6 to 8 in (15 to 20 cm)

Glowing pointed petals twist to reveal darker tips that give the flowers a frosted appearance. Plants have long, strong stems.

**'Bloomquist Paul Jr.'**

**FORM:** Mini Ball

**BLOOM SIZE:** 2 to 3½ in (5 to 9 cm)

Darling cardinal-red flowers are borne on strong stems. Vigorous, productive plants churn out loads of striking rounded blooms.

### 'Comet'

**FORM:** Anemone

**BLOOM SIZE:** 4 to 6 in (10 to 15 cm)

These ruby-red blooms look like fluffy gerbera daisies. Petal backs have an iridescent quality. This great novelty has nice long stems.

### 'Cornel' ♥

**FORM:** Ball

**BLOOM SIZE:** 4 in (10 cm)

Plants produce long, strong stems perfect for cutting, and blooms feature dark cherry-red petals that resemble rich velvet. Flowers are weather resistant and last an extremely long time in the vase. Hands down, it's the best red variety on the market.

### 'Crazy 4 Ieashia'

**FORM:** Formal Decorative

**BLOOM SIZE:** 4 in (10 cm)

Deep scarlet-red blooms have chocolate undertones and fade to copper red as they age. Petal backs are dusted in gold, giving the flowers a glowing quality. Stems are long and strong, and plants are extremely productive.

### 'Creve Coeur'

**FORM:** Semi-Cactus

**BLOOM SIZE:** Over 10 in (25 cm)

Long, strong stems hold up the massive smoky-red blooms of this eye-catching variety. It's a standout in the garden.

### 'Drummer Boy'

**FORM:** Formal Decorative

**BLOOM SIZE:** 8 to 10 in (20 to 25 cm)

Flower heads are perfectly formed on this red beauty, and darker centers add even more dimension to the blooms. Medium-size plants are loaded with flowers.

### 'Gitty Up' ♥

**FORM:** Novelty Fully Double

**BLOOM SIZE:** 3 in (8 cm)

A huge hit with everyone who visits our garden, this eye-catching novelty has fuzzy, cherry-red centers encircled by glowing orange petals. While its blooms look quite delicate, they are surprisingly weather resistant and very productive, and they make a wonderful accent in arrangements.

### 'Jowey Joshua'

**FORM:** Ball

**BLOOM SIZE:** 3½ in (9 cm)

Unusual burnt-red coloring with true gold undertones gives the blooms a metallic appearance. This hardy, vigorous garden addition boasts long, strong stems.

### 'Ken's Choice'

**FORM:** Ball

**BLOOM SIZE:** Over 3½ in (9 cm)

This garden workhorse churns out an abundance of rich ruby-red flowers all season long. The hardy plants produce long-stemmed, weather-resistant flowers that are excellent for cutting.

### 'Lover Boy'

**FORM:** Semi-Cactus

**BLOOM SIZE:** 6 in (15 cm)

The glowing red flowers of this crimson beauty are hard to resist. The subtle blue undertones of the pointed ruby petals and the long stems make them a striking addition to the garden and vase.

### 'Lupin Britain' ♥

**FORM:** Pompon

**BLOOM SIZE:** Up to 2 in (5 cm)

Festive orange-red blooms have slightly lighter petal edges. The small, rounded flowers look like lollipops out in the field.

### 'Ms Scarlett' ♥

**FORM:** Mini Ball

**BLOOM SIZE:** 2 to 3½ in (5 to 9 cm)

Flowers resemble ripe cherries that ride atop long, strong stems. This productive beauty is great for straight bunches and cutting.

### 'Myrtle's Brandy' ♥

**FORM:** Formal Decorative

**BLOOM SIZE:** 4 to 6 in (10 to 15 cm)

This is the most beautiful red-and-white bicolor dahlia we've ever grown. Large plants are covered in striking, slightly nodding blooms. Stems are strong, and flowers hold up well in weather.

**'Sandia Warbonnet'**

FORM: Laciniated

BLOOM SIZE: 6 to 8 in (15 to 20 cm)

This super fun bicolor has smoky-red flowers with white-tipped petals. Shaggy blooms look like costume feathers.

**'Santa Claus'** ♥

FORM: Informal Decorative

BLOOM SIZE: 4 to 6 in (10 to 15 cm)

This fun variety features white petals with reddish-orange centers. Petals curve back and slightly twist, giving the blooms a feathery look. This long-stemmed variety has a delicate appearance, but the flowers hold up well in weather.

**'Spartacus'**

FORM: Informal Decorative

BLOOM SIZE: 9 to 12 in (23 to 30 cm)

This dramatic variety produces an abundance of huge, deep velvet red blooms that consistently win at shows. It's a fabulous tuber producer, too.

**'Tahoma Velvet'** ♥

FORM: Formal Decorative

BLOOM SIZE: 4 to 6 in (10 to 15 cm)

Deep, velvety, rich red flowers sit atop long, strong stems. This is an excellent cut flower and a real standout in the garden.

**'Taum Sauk'**

FORM: Semi-Cactus

BLOOM SIZE: 10+ in (25 cm)

Large red flowers have long pointed petals with black streaking—they remind us of fireworks. This variety has long, strong dark stems.

**'Windmill'**

FORM: Single

BLOOM SIZE: 4 in (10 cm)

This campy novelty should be called Peppermint Taffy with its curled red-and-white petals, separated and spaced around a golden center. Extremely long lasting in the vase.

# MAROON/BLACK

Filled with drama, intrigue, and mystery, this group
has a surprisingly broad variety of nuanced color,
from jet black to rich maroon, with subtle cool
purple or warm red undertones.

### 'Black Narcissus' ♥

**FORM**: Laciniated

**BLOOM SIZE**: 6 to 8 in (15 to 20 cm)

This spiky black treasure was a gift from a dear friend, and in addition to churning out buckets of textural flowers, it adds drama to the garden and the vase. Flowers are a bit weather sensitive, so harvest before fully mature and let them open completely indoors.

### 'Bumble Rumble'

**FORM**: Collarette

**BLOOM SIZE**: 3 to 4 in (8 to 10 cm)

Reminding us of a clown's collar, outer white petals with raspberry and purple veining surround a ring of shorter white petals. This variety is a real conversation starter, and bees love them as well.

### 'Chimacum Night'

**FORM**: Formal Decorative

**BLOOM SIZE**: Up to 4 in (10 cm)

The beautiful mahogany flowers of this treasure bloom abundantly all season long. The flowers ride atop tall, strong stems. This variety is both striking and productive.

### 'Chimacum Troy'

**FORM**: Mini Ball

**BLOOM SIZE**: 2 to 3½ in (5 to 9 cm)

This garden workhorse boasts long, strong stems and perfect uniform blooms. Flowers are the color of red wine. It's a consistent show winner.

### 'Diva' ♥

**FORM**: Formal Decorative

**BLOOM SIZE**: 6 in (15 cm)

These towering plants produce sangria-toned blooms with perfectly formed petals. This color is absolutely stunning and glows in the garden.

### 'Glenplace'

**FORM**: Pompon

**BLOOM SIZE**: Up to 2 in (5 cm)

A longtime staple in our cutting garden, these rich merlot blooms are weather resistant and great for straight bunches.

### 'Hollyhill Black Beauty'

**FORM**: Informal Decorative

**BLOOM SIZE**: 4 to 6 in (10 to 15 cm)

Tall, vigorous plants produce an abundance of long-stemmed, velvety black blooms. It's excellent for cutting.

### 'Inego'

**FORM**: Formal Decorative

**BLOOM SIZE**: Up to 4 in (10 cm)

Flowers are a striking color, starting out rich maroon with outer petals aging to magenta. This tall, vigorous grower has long, strong stems.

### 'Irish Blackheart'

**FORM**: Stellar

**BLOOM SIZE**: 4 to 6 in (10 to 15 cm)

This is one of the best dark bicolor varieties I've grown. The maroon-black petals have bright white centers and tips, giving the blooms a distinctive appearance.

### 'Ivanetti' ♥

**FORM**: Ball

**BLOOM SIZE**: Over 3½ in (9 cm)

A top pick in this color family, plants are loaded with rich berry-hued blooms all season. Long, strong stems and weather-resistant flowers make them a fantastic cut.

### 'Jessie G.'

**FORM**: Ball

**BLOOM SIZE**: Over 3½ in (9 cm)

This productive burgundy gem has strong stems that are easy to arrange with. Large flowers are excellent for cutting and market bunches.

### 'Jowey Marilyn'

**FORM**: Formal Decorative

**BLOOM SIZE**: 4 to 6 in (10 to 15 cm)

The wine-colored flowers are borne on long, strong stems. Petal tips have the lightest silver dusting, giving the flowers a metallic quality.

### 'Jowey Mirella' ♥

**FORM:** Ball

**BLOOM SIZE:** 3 to 4 in (8 to 10 cm)

Both hardy and productive, this dramatic beauty blooms in abundance all season. Tall, strong stems make it an excellent cut flower.

### 'Jump Start' ♥

**FORM:** Anemone

**BLOOM SIZE:** 4 in (10 cm)

Chocolaty maroon flowers have a tufted dark center encircled by a single row of petals. It reminds us of chocolate cosmos. Plants are not super tall but are very productive and have a branching habit.

### 'Karma Choc'

**FORM:** Waterlily

**BLOOM SIZE:** 4 to 5 in (10 to 13 cm)

These velvety merlot-hued blooms have slightly lighter outer petals. Flowers on this low grower actually smell like chocolate and are worth growing for the scent alone.

### 'Moor Place'

**FORM:** Pompon

**BLOOM SIZE:** Up to 2 in (5 cm)

This adorable button dahlia is as hard-working as it is productive. Each plant is smothered in miniature deep maroon flowers all season. Long, strong stems and weather resistance make it an excellent variety for cutting.

### 'Natal' ♥

**FORM:** Mini Ball

**BLOOM SIZE:** 3 in (8 cm)

One of the most consistent and productive dark-flowered varieties we grow, this workhorse churns out armloads of weather-resistant blooms on long, strong stems. Flowers are perfect in hand-tied bouquets.

### 'Poodle Skirt' ♥

**FORM:** Novelty Fully Double

**BLOOM SIZE:** 3 in (8 cm)

This low grower is covered by the most adorable dusty-mauve blooms. Outer petals are extremely reflexed and surround a fluffy center. Most people wouldn't guess it's a dahlia.

### 'Rocco'

**FORM:** Pompon

**BLOOM SIZE:** Up to 2 in (5 cm)

This cute miniature variety is loaded with adorable boysenberry-colored flowers, each packed with hundreds of inward-curving petals. While plants are on the shorter side, this abundant flowering variety is great for cutting.

### 'Rock Star'

**FORM:** Anemone

**BLOOM SIZE:** 3 in (8 cm)

Anemone-shaped blooms are beautiful deep cranberry with fuzzy cushion centers. This adorable garden workhorse boasts long stems.

### 'Soulman' ♥

**FORM:** Anemone

**BLOOM SIZE:** 4 to 6 in (10 to 15 cm)

Rich velvety maroon flowers with dark centers have many layers of petals, with back petals that reflex slightly. As flowers fade, tips turn crimson. Plants have ferny foliage.

### 'Tam Tam'

**FORM:** Ball

**BLOOM SIZE:** Over 3½ in (9 cm)

This hardworking bloomer is smothered in reddish-black flowers all season long. Because of its bloom size, striking color, and long stems, it makes a wonderful cut flower.

### 'Tartan'

**FORM:** Informal Decorative

**BLOOM SIZE:** 7 to 8 in (18 to 20 cm)

Twirling petals are a deep maroon with white tips. Long, strong stems carry large flower heads with varied coloring—no two flowers are the same.

### 'Verrone's Obsidian'

**FORM:** Orchid

**BLOOM SIZE:** 4 to 5 in (10 to 13 cm)

Medium-size plants are covered in the deepest black star-shaped flowers. Petals curl in slightly, giving them a striking look. Pick when young, as bees love them.

# RESOURCES

**FLORET DAHLIA GUIDE**

www.floretflowers.com/dahlias

An up-to-date list of our favorite dahlia suppliers along with other resources to help you grow and learn even more about these incredible flowers.

## SUPPLIES

**A.M. Leonard**
www.amleo.com
Tools and supplies, including Atlas 370 nitrile gloves, pitchforks, and loppers.

**Amazon**
www.amazon.com
Rooting hormone (I like Clonex Rooting Gel), flower food, Atlas 370 nitrile gloves, and metal plant tags.

**Dripworks**
www.dripworks.com
Irrigation supplies, including drip tape and earth staples.

**Farmhouse Pottery**
www.farmhousepottery.com
Great source for unique, handcrafted pottery vases.

**Floral Supply Syndicate**
www.fss.com
This national chain carries flower food as well as a wide range of vases and floral design supplies.

**Floret Farm**
www.floretflowers.com
Our online shop stocks seeds for many of the arrangements in this book, as well as flower arranging supplies, gardening tools, and bulbs.

**Frances Palmer Pottery**
www.francespalmerpottery.com
My favorite source for handmade ceramic vases.

**Growing Solutions**
www.growingsolutions.com
Compost tea supplies and equipment.

**Johnny's Selected Seeds**
www.johnnyseeds.com
A wonderful mail-order source for seed trays, heat mats, specialty tools, natural fertilizers, and insect repellents.

**Paw Paw Everlast Label Company**
www.everlastlabel.com
A great source for metal plant tags in bulk and marking products.

**UMass Soil and Plant Tissue Testing Lab**
http://ag.umass.edu/services/soil-plant-nutrient-testing-laboratory
Mail-order soil testing laboratory.

219

# ACKNOWLEDGMENTS

### ERIN BENZAKEIN

This book was a true labor of love, and there were so many amazing people who had a hand in making it. Chris, you really outdid yourself with all of the beautiful images on these pages. Thank you for always supporting my big ideas and helping me bring this book to life. I'm so grateful that we get to do this crazy life together! Elora and Jasper, thank you for all of the evenings you spent helping Dad and me photograph the armloads for this book. I know it wasn't the most exciting way to spend your summer evenings, but you sure didn't let it show. You guys are such troopers! Mom, thank you for all of the early morning coaching sessions and encouraging me to keep going, even when I wanted to quit. Jill, I couldn't ask for a better partner in crime. Writing this book was as fun as it was challenging. Thank you for keeping us laughing the entire time and believing we could do it, even when my faith wavered. Your friendship and the work we get to do together means the world to me. Julie, you have been the calm in the storm. Thank you for your steady support and incredible patience throughout the entire process. Without you, this book wouldn't exist. A huge thank you to Martha Stewart and Joanna Gaines for your generous support. Nina, I'm so happy you made it home in time to help with the last of the bouquets! Having your magic on the pages is everything. Marryn Mathis, massive thanks for your dedication, enthusiasm, and attention to detail on this incredibly complex project. The rainbow show field was a masterpiece and we couldn't have done this without all of your amazing help. Kristine Albrecht, I'm so grateful for the time you spent with me in the early stages of mapping out this book. Your wise words and generous information were priceless. Ken Greenway, the dahlia breeding world is a very secretive one. Thank you for being so willing to answer a million of my novice questions over the years and for sharing some of your coveted seeds with me. Jan Johnson, I don't think you know just how much your generosity has impacted my life. Thank you for giving me my very first dahlia tubers and lighting a fire in me for gardening and sharing that still burns bright. To our farm crew, I so appreciate all of the love and care you poured into growing the beautiful flowers on these pages. Susan Studer King, thank you for helping spread this book far and wide. And to the ladies of Team Floret, thanks so much for holding down the fort while we squirreled away and made this book, in the middle of our busiest season yet. Your encouragement and patience made all the difference. Leslie Jonath and Leslie Stoker, thanks for your support and always calmly helping me navigate the tricky waters of publishing. You two are the best. Massive thanks to Rachel Hiles and the Chronicle Books team for taking a chance on an obscure gardening book and believing that there are enough obsessed dahlia lovers out there to make it a success. Ashley Lima, thank you for lending your amazing design eye to this project and making it our most beautiful book yet. Thank you, thank you, thank you, to the mystery person who sent me a box of 'Castle Drive' tubers all those years ago. They continue to multiply and bless countless people each year. Lastly, to all of the Floret fans, readers, and customers, thank you for your enthusiasm about this book. I hope you love it as much as I loved making it for you!

### JILL JORGENSEN

Writing this book was beautiful, fun, and hard. I'm so glad I was able to play a part in it. Erin, writing with you is the best. I love it when we figure out how to explain complicated things and make them beautiful. It's the honor of a lifetime. Chris, thank you for your positive attitude, for always having fresh batteries at the ready, and for your beautiful photos (you even made crown gall look good). Julie Chai, thank you for wrapping your mind around this one and staying on it, and for your commitment to making the best book—every time. Ashley Lima, thank you for your patience and thoughtfulness as we worked to get this book just right; it's an absolute work of art. Marryn Mathis, your superhuman project management skills were a sight to behold, and you would totally kick our asses at dahlia memory. Thank you for keeping us so organized. Susan Studer King, thank you for launching yet another Floret title out into the world. To Team Floret, thank you for planting, tending, harvesting, digging, and dividing everything that went into this book. Your contributions to our creative projects are greater than you know. Big thanks to my husband, Joel, and kids, Cora and Felix, for acting relatively unfazed while I wrote two books in 1 year. I appreciate the slack, and please know that what I'm trying to show you is that if you focus and work really hard, seemingly impossible things can happen. Love and thanks to my family for their support and excitement about my work. To all the Floret fans, we set out to make the most informative and beautiful book we could. I hope it inspires you for years to come. So often in the process of creating this book, I thought of my grandpa George and the vision of him standing with pride in front of his prized (albeit highlighter yellow) dinner plate dahlias. I sure wish we could talk about them now.

## JULIE CHAI

Erin and Jill, you are superwomen. Taking on this book when we did couldn't have been more challenging, but in spite of all the twists and turns and tight scheduling, we pulled it off, thanks to your unwavering commitment to getting it done, and getting it done right. Erin, I am grateful that

you dream big, trust the power of possibility, and let your love of flowers, and the beauty they bring to others, drive you. Jill, you are the definition of grace under pressure. Thank you for being so on top of the details and keeping the mood light at every turn. Chris, your photos are magic. Thank you for taking such care in capturing the gorgeousness and vitality of the farm to bring all the words to life. Leslie Jonath, thank you for your wisdom and constant encouragement and for always seeing the big picture. Leslie Stoker, thank you for your steadiness, support, and expert guidance. Rachel Hiles and the Chronicle Books team, thank you for your collaborative spirit and being so supportive of our vision for this book. Ashley Lima, thank you for being so enthusiastic about how we wanted to tell this story and translating our mound of notes, sketches, and ideas into these stunning pages. Mom and Dad, thank you for always cheering me on and for showing me exactly what it looks like to follow your own path and live the life you want. My George and Ellis, thank you for taking it in stride when this book pulled me away during evenings, weekends, and family trips; for being excited about everything we grow; and for appreciating my novice bouquet-making skills. I'm grateful for you every single day.

## CHRIS BENZAKEIN

Erin, love, I'm so grateful to be together and create something that will live on throughout time. Elora and Jasper, thanks for cheering us on, even when we weren't as available as we wanted to be. Chérie, thank you for only speaking to my best self. Dad, thanks for reminding me about what's really important. Jill, thank you for always being there to bounce ideas back and forth with Erin and for putting such a fun spin on really hard things. Julie, thanks a million for keeping this project so dialed. And big thanks to Team Floret for helping everything run smoothly while we had our heads down on this book.

221

# INDEX

222

# INDEX

out how to explain complicated things and make them beautiful. It's the honor of a lifetime. Chris, thank you for your positive attitude, for always having fresh batteries at the ready, and for your beautiful photos (you even made crown gall look good). Julie Chai, thank you for wrapping your mind around this one and staying on it, and for your commitment to making the best book—every time. Ashley Lima, thank you for your patience and thoughtfulness as we worked to get this book just right; it's an absolute work of art. Marryn Mathis, your superhuman project management skills were a sight to behold, and you would totally kick our asses at dahlia memory. Thank you for keeping us so organized. Susan Studer King, thank you for launching yet another Floret title out into the world. To Team Floret, thank you for planting, tending, harvesting, digging, and dividing everything that went into this book. Your contributions to our creative projects are greater than you know. Big thanks to my husband, Joel, and kids, Cora and Felix, for acting relatively unfazed while I wrote two books in 1 year. I appreciate the slack, and please know that what I'm trying to show you is that if you focus and work really hard, seemingly impossible things can happen. Love and thanks to my family for their support and excitement about my work. To all the Floret fans, we set out to make the most informative and beautiful book we could. I hope it inspires you for years to come. So often in the process of creating this book, I thought of my grandpa George and the vision of him standing with pride in front of his prized (albeit highlighter yellow) dinner plate dahlias. I sure wish we could talk about them now.

### JULIE CHAI

Erin and Jill, you are superwomen. Taking on this book when we did couldn't have been more challenging, but in spite of all the twists and turns and tight scheduling, we pulled it off, thanks to your unwavering commitment to getting it done, and getting it done right. Erin, I am grateful that you dream big, trust the power of possibility, and let your love of flowers, and the beauty they bring to others, drive you. Jill, you are the definition of grace under pressure. Thank you for being so on top of the details and keeping the mood light at every turn. Chris, your photos are magic. Thank you for taking such care in capturing the gorgeousness and vitality of the farm to bring all the words to life. Leslie Jonath, thank you for your wisdom and constant encouragement and for always seeing the big picture. Leslie Stoker, thank you for your steadiness, support, and expert guidance. Rachel Hiles and the Chronicle Books team, thank you for your collaborative spirit and being so supportive of our vision for this book. Ashley Lima, thank you for being so enthusiastic about how we wanted to tell this story and translating our mound of notes, sketches, and ideas into these stunning pages. Mom and Dad, thank you for always cheering me on and for showing me exactly what it looks like to follow your own path and live the life you want. My George and Ellis, thank you for taking it in stride when this book pulled me away during evenings, weekends, and family trips; for being excited about everything we grow; and for appreciating my novice bouquet-making skills. I'm grateful for you every single day.

### CHRIS BENZAKEIN

Erin, love, I'm so grateful to be together and create something that will live on throughout time. Elora and Jasper, thanks for cheering us on, even when we weren't as available as we wanted to be. Chérie, thank you for only speaking to my best self. Dad, thanks for reminding me about what's really important. Jill, thank you for always being there to bounce ideas back and forth with Erin and for putting such a fun spin on really hard things. Julie, thanks a million for keeping this project so dialed. And big thanks to Team Floret for helping everything run smoothly while we had our heads down on this book.

221

223